THE ELEMENTS

Ingredients of the Universe

Written and illustrated by
Ellen Johnston McHenry

ISBN: 978-0-9825377-1-8

Ellen McHenry's Basement Workshop
www.ellenjmchenry.com
State College, PA 16801
ejm.basementworkshop@gmail.com

Also by this author:

Carbon Chemistry
Botany in 8 Lessons
Cells
The Brain
Protozoa: A Poseidon Adventure
Mapping the World with Art
Rocks and Dirt

Periodic Table of Elements

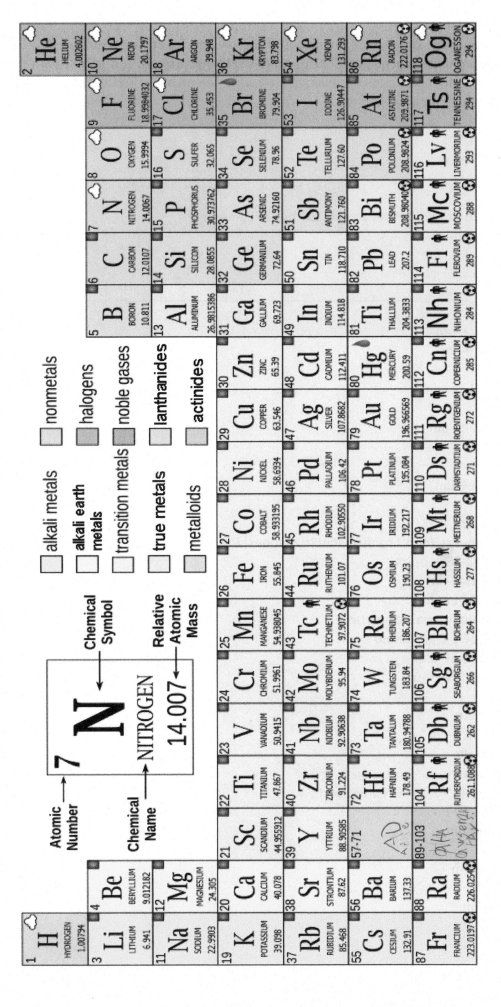

In your reading, you may come across the names of these elements and be unsure of how to pronounce them. This pronunciation guide will help you to say them correctly. The syllable with the capital letters is the one that you give emphasis. (For example, the word "element" would be "EL-eh-ment.") Turn back to this page whenever you need to!

Actinium: act-IN-ee-um
Americium: am-air-ISH-ee-um
Antimony: AN-teh-mo-nee
Arsenic: AR-sen-ick
Berkelium: BERK-lee-um (*though many people say* ber-KEEL-ee-um)
Beryllium: beh-RILL-ee-um
Boron: BORE-on
Cerium: SEER-ee-um
Cesium: SEE-zee-um
Curium: KYOOR-ee-um
Dysprosium: dis-PRO-zee-um
Europium: yoo-ROPE-ee-um
Fluorine: FLOR-een
Gadolinium: GAD-o-LIN-ee-um
Gallium: GAL-ee-um
Germanium: jer-MANE-ee-um
Iridium: er-RID-ee-um
Krypton: KRIP-tohn
Lawrencium: lore-EN-see-um
Lithium: LITH-ee-um
Lutetium: loo-TEE-she-um
Manganese: MANG-gan-eez (*don't confuse it with magnesium!*)
Mendelevium: men-dell-EE-vee-um
Molybdenum: moll-IB-den-um
Neodymium: NEE-o-DIM-ee-um
Palladium: pal-AID-ee-um (*or* pal-AD-ee-um)
Praseodymium: PRAZ-ee-o-DIM-ee-um
Promethium: pro-MEE-thee-um
Protactinium: PRO-tack-TIN-ee-um
Rhodium: ROE-dee-um
Rubidium: roo-BID-ee-um
Ruthenium: roo-THEE-nee-um
Samarium: sam-AIR-ee-um
Selenium: seh-LEEN-ee-um
Strontium: STRON-tee-um (*or* STRON-shee-um)
Technetium: teck-NEE-she-um
Tellurium: tell-LOOR-ee-um
Thulium: THOO-lee-um
Vanadium: van-AY-dee-um
Xenon: ZEE-non
Ytterbium: i-TER-bee-um
Yttrium: IT-ree-um

CHAPTER 1: WHAT IS AN ELEMENT?

Do you ever help bake things like cookies, cakes, biscuits, or bread? If so, you may have noticed that all baked goods are made from basically the same ingredients: flour, sugar, salt, eggs, butter, vegetable oil, baking powder, yeast and flavorings. The ingredients can be the same, or at least very similar, yet you have no problem telling the difference in taste and texture between pancakes and donuts, or biscuits and bread.

Even though these foods contain many of the same ingredients, the ingredients are used in different proportions. Cookies, for example, have lots of butter and sugar and not too much flour. Biscuits have less sugar than cookies do, and contain no eggs. Bread is mostly flour, with only a small amount of sugar and butter or oil (and some yeast to make it rise). Some recipes call for flavorings such as cinnamon, chocolate or lemon. The same ingredients in your kitchen can be used in many different ways to make many different foods.

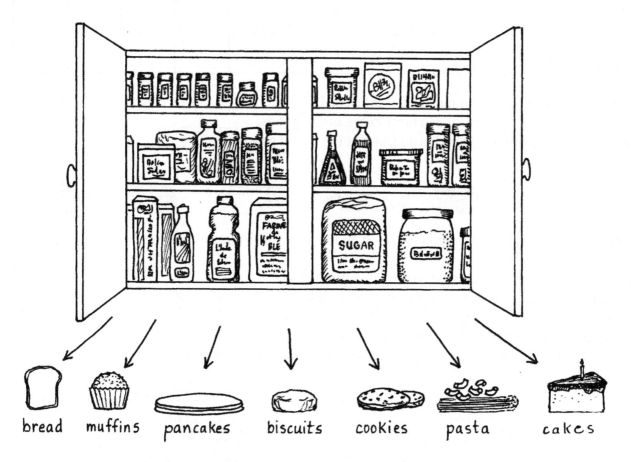

bread   muffins   pancakes   biscuits   cookies   pasta   cakes

All of these foods can be made from the ingredients in your cupboard. The reason they are different is that they have more of some things and less of others. Just a pinch of flavoring or spice can change one recipe into another. It doesn't take thousands or millions of ingredients to make a wide variety of recipes. Most of us have less than 100 ingredients in our cupboards, yet we can use them to make just about any recipe we find in a cookbook.

Activity 1.1

Use a cookbook to find the information for this activity, or ask an adult who knows a lot about cooking. For each baked good, put check marks in the boxes, showing what ingredients it contains. You are free to choose any recipes you like. (Flour means any kind, including gluten-free types. You may also cross out banana or chocolate put in something like blueberries or nuts instead.)

	flour	sugar	oil or eggs	milk or butter	water	yeast	baking powder	vanilla	banana	chocolate or other flavor
BREAD										
COOKIES										
BISCUITS										
PANCAKES										
CAKE										
BANANA MUFFINS										

Name an ingredient that is found in all of the baked goods:_____

Name an ingredient that is found in most of the baked goods: _____

Name an ingredient that is found in only one of the baked goods: _____

Activity 1.2

Think about cookies (tough assignment, eh?) and answer these questions:

1) How would a cookie change if you put it in the freezer?
2) How would a cookie change if you let it sit out somewhere for a week?
3) How would a cookie change if you put it in a glass of water?
4) Do these changes mean that the recipe changed?
5) Do other factors, not just the recipes, contribute to the quality of foods?

Okay, so baked goods are made out of ingredients.

Right.

But do those ingredients have ingredients? Like, what is flour made of?

Let's read on...

We know that baked goods are made of ingredients. But what are *ingredients* made of? What is flour made of? What is water? What is oil? These baking ingredients are made of chemical ingredients called **elements**. The chemical elements are the most basic ingredients of all. They are the things that everything else is made of. There are a little over 100 chemical elements, and if we could put a sample of each into a little bottle or box, we'd have sort of a "kitchen cupboard of the universe."

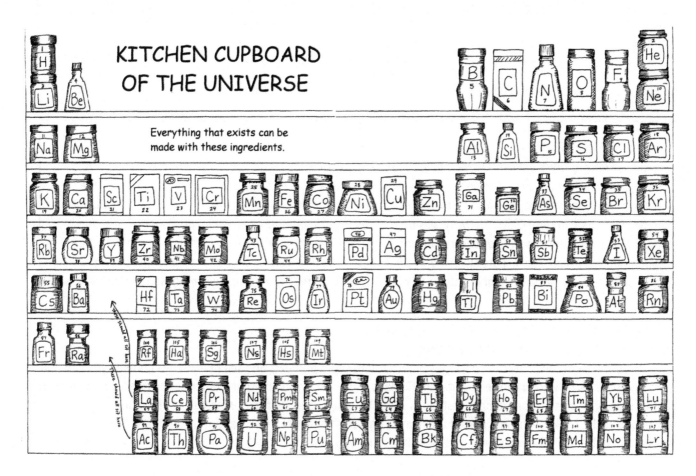

KITCHEN CUPBOARD OF THE UNIVERSE

Everything that exists can be made with these ingredients.

These are the ingredients that make up anything you can think of: plants, animals, rocks, plastic, metal, fuel, fabric, computers, food, water, air, garbage... anything! Your body is made of these elements, too. You are a "recipe" of these chemical ingredients.

Some of these chemical elements are very common and are found in practically everything, just like flour is found in so many baked goods. You may already be familiar with the names of some of these common elements: hydrogen, carbon, nitrogen, oxygen and silicon. These five elements account for most of the matter (stuff) in the universe! Other elements are less common and have names you've never heard of, such as osmium or ruthenium. These uncommon elements are a bit like the spices lurking at the back of your cupboard— the ones you use only once in a while, such as dill weed or coriander.

Isn't it great to find out that you already know some of these elements? Another chemical element you are already familiar with is helium. You've known about that one since you were old enough to hold a balloon. You just didn't know it was one of the basic ingredients of the universe. You probably know quite a few more, too, like gold, silver, lead, iron, copper, nickel, and aluminum. How many others do you know?

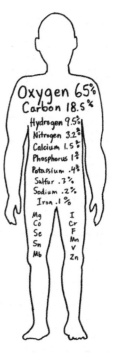

Oxygen 65%
Carbon 18.5%
Hydrogen 9.5%
Nitrogen 3.2%
Calcium 1.5%
Phosphorus 1%
Potassium .4%
Sulfur .3%
Sodium .2%
Iron .1%

Mg   I
Co   Cr
Se   F
Sn   Mn
Mb   V
      Zn

Activity 1.3   Elements you already know

How many of these elements do you recognize? Circle any name that you have heard of, even if you don't know exactly what it is. (This is not a complete list of all elements, only about half of them.)

hydrogen	helium	lithium	boron	carbon
nitrogen	oxygen	fluorine	neon	sodium
magnesium	aluminum	silicon	phosphorus	sulfur
chlorine	potassium	calcium	manganese	titanium
chromium	iron	cobalt	nickel	copper
zinc	lead	silver	gold	platinum
mercury	arsenic	selenium	tin	radon
uranium	plutonium	iodine	zirconium	tungsten

ELEMENTS IN HISTORY:

Some of these elements were familiar to ancient peoples. Silver and gold, for example, have been used for thousands of years. The ancients also knew about iron, tin, lead, copper, sulfur, and mercury. (They didn't understand what a chemical element was, however, and they also considered fire, water, earth and air to be elements.) In the 1800s, electricity was used to discover magnesium, potassium and sodium. Also in the 1800s, new elements were discovered in mines. In the 1900s, radioactive elements such as uranium and plutonium were discovered. (They were named in honor of the discovery of Uranus and Pluto just a few years previously.) Elements with numbers above 92 did not exist until they were artificially made in the late 1900s.

Activity 1.4   A scavenger hunt for elements

Read the labels on some food packages or other household products and see how many elements you can find. (Pet foods are especially good choices.) Put a check mark in the box if you find that element. The names of the elements might be slightly disguised. For example, instead of sulfur you might see "sulfite," or instead of phosphorus you might see "phosphoric acid." Look for the first part of the names, and don't worry too much about the endings. The three empty spaces at the bottom are for you to add other elements that you find.

You choose three more: B D

	cereal	medicine or toothpaste	bread	Shampoo	periodic table of elements	food
calcium	✓	✓	✓		✓	
carbon	✓	✓			✓	
chlorine					✓	
copper		✓			✓	
fluorine		✓			✓	
iodine					✓	
iron	✓		✓		✓	
phosphorus		✓			✓	✓
potassium					✓	✓
magnesium					✓	
zinc			✓		✓	
Sodium	✓	✓		✓	✓	✓
silicon		✓			✓	
Sulfure		✓		✓	✓	

So what are *ingredients* made of? Is there a recipe to make salt or sugar? Yes, there is! The ingredients are the chemical elements and the recipes are called formulas. For example, to make salt, you need two chemical elements: sodium and chlorine. If you combine these two elements together, you will get table salt. The recipe for sugar calls for three elements: carbon, hydrogen, and oxygen. Some chemical recipes, like sugar and salt, are fairly simple. Other materials have recipes that are extremely complicated. Livings things, such as plants and animals, are also made of chemical elements but are mixtures of so many different substances that you really can't come up with a recipe for them.

A cooking recipe looks like this:

Sugar cookies:
2 cups flour   1 egg
1/2 cup sugar   1 teaspoon vanilla
1/2 cup butter   1/2 teaspoon baking soda

A chemical recipe looks like this:

glucose sugar = $C_6H_{12}O_6$

The letters are abbreviations, or **symbols**, for elements. C stands for carbon, H stands for hydrogen, and O stands for oxygen. The numbers below the letters tell you how many of each atom go into the recipe. This recipe calls for 6 atoms of carbon, 12 atoms of hydrogen and 6 atoms of oxygen. Just like with a cooking recipe, you can make a small, medium, or large batch. Theoretically, you could make a batch as small as a few molecules or large enough to fill a dump truck. As long as you keep the number of atoms in the ratio 6, 12, 6, you will get glucose sugar.

Let's look at the recipe for water:

water = H_2O

The elements in this recipe are similar to the one for glucose sugar, except that there is no carbon. You will need just hydrogen and oxygen. How much of each? There are 2 hydrogen atoms and... but there is no number after the O. Now what? If you don't see a number, it means there is only one. Scientists decided a long time ago that it was too much work to put in all the 1's in the recipes, so they agreed to just leave them out. If you don't see a number after the letter, that means there is only one. (You could think of the 1's as being invisible.)

We'll need 2 atoms of hydrogen for every 1 atom of oxygen. How much of the recipe will you make? A glass of water, or enough to fill a swimming pool? (The fascinating thing about this recipe is that when you combine two gases you get a liquid. And if you break water apart, you get two gases again.)

What about the recipe for salt?

table salt = NaCl

We don't see any numbers here at all. That means one atom of each. What are the ingredients? **Na** is the letter symbol for sodium (which used to be called natrium) and **Cl** is the abbreviation for chlorine (yes, chlorine goes in your pool, too, but it is also in salt).

Let's look at the recipe for baking soda:

baking soda = NaHCO$_3$

That's 1 atom of sodium, 1 atom of hydrogen, 1 atom of carbon, and 3 atoms of oxygen. Those are all the same ingredients we just used to make salt and sugar, but if you combine them in this proportion you will make baking soda. (Baking soda's job in kitchen recipes is to make things "puff up" in the oven.)

What else can we make with chemical elements? Here are some recipes that aren't edible:

sand: SiO$_2$ **Epsom salt: MgSO$_4$**   **gold: Au**   **pyrite ("fool's gold"): FeS$_2$**

We have some new elements in these recipes. **Si** is silicon, **Mg** is magnesium, **Fe** is iron, **S** is sulfur, and **Au** is gold. You can see that the recipe for gold is pretty simple—it's just the element gold with nothing added. Until the 1700s, scientists did not have a clear idea about the chemical elements. They thought that perhaps it was possible to change other materials into gold. You can see why fool's gold can never become real gold. Iron and sulfur will always be iron and sulfur.

Here is a really long recipe:

a mineral called Vesuvianite: Ca$_{10}$Mg$_2$Al$_4$(SiO$_4$)$_5$(Si$_2$O$_7$)$_2$(OH)$_4$

Wow! We won't be cooking up any of that!

* *

Activity 1.5   Making larger batches

Recipes can be doubled, tripled, or cut in half, depending upon how much of the product you want to make. See if you can figure out the answers to these recipe questions.

(Note: We're just using an imaginary "scoop" that accurately counts the atoms for us. In real life, measuring elements and mixing them requires special equipment and more difficult math.)

1) The recipe for the mineral calcite is CaCO$_3$. If we use 2 "scoops" of Ca (calcium), how many "scoops" of the other ingredients will we need? C = _____   O = _____

2) The recipe for the mineral called cinnabar (sounds delicious, but it's poisonous) is HgS. If we make a batch of cinnabar using 3 "scoops" of Hg (mercury), how many "scoops" of S (sulfur) will we need? _____

3) You are a practical joker and want to make a batch of fool's gold to trick a friend. The recipe for fool's gold is FeS$_2$. If you use 4 "scoops" of S (sulfur) how many "scoops" of Fe (iron) will you need? _____

4) A mineral gemstone called zircon can sometimes resemble a diamond. The recipe to make zircon is ZrSiO$_4$. If you use 2 "scoops" of Zr (zirconium), how many "scoops" of the other ingredients will you need? Si = _____   O = _____

Activity 1.6

See if you can match the element with the meaning of its name.

1) Named after Alfred Nobel, inventor of dynamite and founder of the Nobel Prizes _____

2) Named after Vanadis, a goddess from Scandinavian mythology _____

3) Named after Johan Gadolin, a Finnish chemist _____

4) Named after Poland, the country in which famous chemist Marie Curie was born _____

5) Named after Albert Einstein _____

6) Named after the city of Berkeley, California _____

7) Named to honor our planet, Earth, but using the Greek word for Earth: "Tellus" _____

8) Named for the area of Europe called Scandinavia (Norway, Finland, Sweden, Denmark) _____

9) Named for the Swedish town of Ytterby _____

10) Named for Niobe, a goddess in Greek mythology who was the daughter of Tantalus _____

11) Named for Tinia, a mythological god of the Etruscans (in the area we now call Italy) _____

12) Named for Stockholm, Sweden _____

13) Named in honor of the discovery of the planet Neptune _____

14) Named in honor of Marie and Pierre Curie, who discovered radium and polonium _____

15) Named after the Roman messenger god, Mercury, who had wings on his feet _____

16) Named after the Greek god Tantalus (father of Niobe) _____

17) Named in honor of the discovery of the asteroid Ceres _____

18) Named after France, but using its ancient name, Gall _____

19) Named after the moon, but using the Greek word for moon, "selene" _____

20) Named for its really bad smell, using the Greek word "bromos" which means "stench" _____

21) Named after the Latin word for rainbow, "iris," because it forms salts of various colors _____

22) Named after Thor, the Norse god of thunder _____

23) The name comes from the German word "Kupfernickel," meaning "Satan's copper" _____

24) The name comes from the German "Kobald," a mythological gnome who lived in mines _____

25) Named for its color, yellowish-green, using the Greek word for this color: "chloros" _____

THE POSSIBLE ANSWERS: (If you need help with pronunciation, use the key before page 1.)

berkelium, bromine, cobalt, cerium, chlorine, curium, einsteinium, gadolinium, gallium, holmium, iridium, mercury, neptunium, nickel, niobium, nobelium, polonium, scandium, selenium, tantalum, tellurium, thorium, tin, vanadium, ytterbium

Just use logical reasoning to figure them out!

Activity 1.7   "The Chemical Compounds Song"

Here is a very silly song about chemical recipes. The audio tracks for this song can be found at www.ellenjmchenry.com/audio-tracks-for-the-elements (or in the zip file if you have the digital download). There are two versions of this song. The first one has the words so you can learn how they match the tune. The second version is accompaniment-only so you can sing it yourself. When singing it becomes easy, try it as a hand-clap game, like "Miss Merry Mack" or "Down, Down Baby." You don't even need the music if you use it as a hand-clap game. (Also, there is a music video of this song posted on the YouTube playlist mentioned at the top of page 17.)

The Chemical Compounds Song

Today was Mama's birthday; I tried to bake a cake.
I didn't use a recipe, that was my first mistake!

I put in lots of H_2O, 3 cups NaCl,
Some $NaHCO_3$, and other things as well.

I poured it in a non-stick pan (Teflon, C_2F_4)
I popped it in the oven (it cooks with CH_4).

I set the oven way too hot, the cake got black and charred.
Oh, why did I make birthday cake? I should have bought a card!

I had to clean and scrub the pan, so Mom would never know.
First I tried to bleach the pan with NaClO.

I needed something stronger, so I tried some HCl.
I added grit, SiO_2, and FeO, as well.

Then something awful happened, I'll never know just why.
I woke up in the hospital with stitches near my eye!

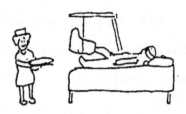

My leg was in a plaster cast of $CaSO_4$.
The nurse brought $Mg(OH)_2$ and $MgSO_4$.

Next year for Mama's birthday, I'll buy a cake, instead,
'Cause if I tried to bake again, I think I'd end up dead!

H_2O = water
NaCl = salt
$NaHCO_3$ = baking soda
C_2F_4 = Teflon
CH_4 = natural gas
NaClO = bleach

HCl = hydrochloric acid
SiO_2 = sand
FeO = a type of rust [or FeO(OH) to be more accurate] *
$CaSO_4$ = plaster
$Mg(OH)_2$ = milk of magnesia (good for intestines)
$MgSO_4$ = Epsom salt (good for skin)

* Rust is complicated. More commonly it is written as $Fe_2O_3(OH)$ or $Fe_2O_3.nH_2O$. But those don't fit the rhyme.

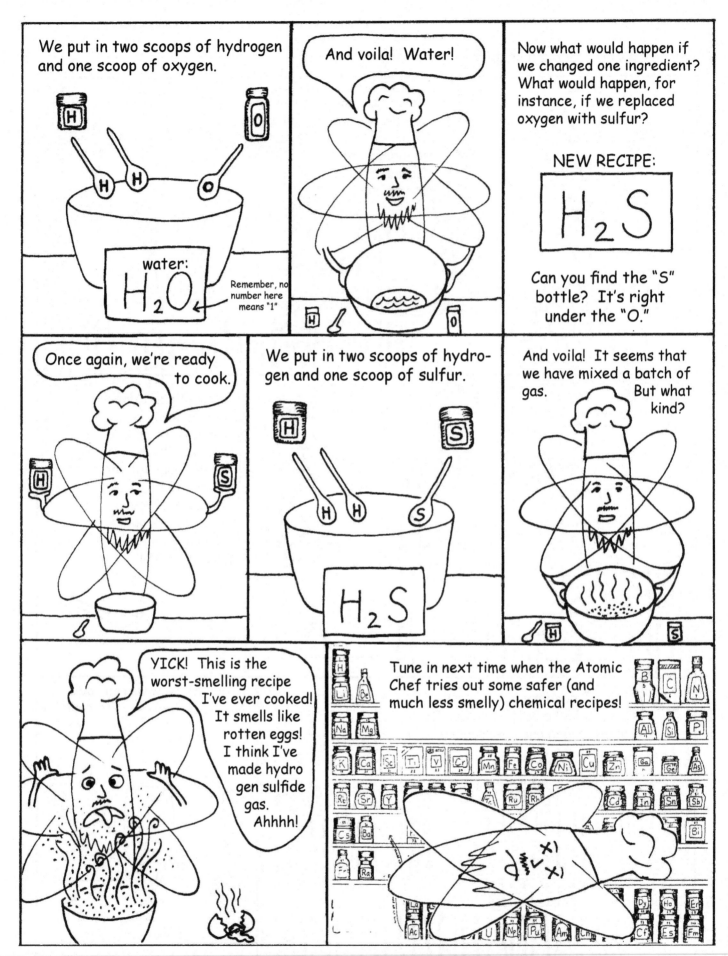

CHAPTER 2: THE PERIODIC TABLE

Have you ever read stories from medieval times where a person called an "alchemist" tried to make gold? The alchemists were part scientist and part magician, and although they experimented with other forms of chemistry, they are famous for trying to make gold. They boiled up mixtures of every substance they could find: copper, tin, lead, iron, coal, silver, mercury, unusual rocks, gold-colored minerals, medicinal plants, parts of animals, and anything else they could think of. They even said magic spells over their boiling pots, but they never produced a single drop of gold.

What the alchemists did not know is that gold is one of the basic ingredients in the Kitchen Cupboard of the Universe. They thought gold had a recipe like water or salt or sugar. But you can't make gold. It's a basic ingredient that comes naturally in the Earth. The letter symbol for gold is **Au**. Ancient peoples called gold by the name "aurum" and that is where we get the letters "Au." Can you find gold in the Kitchen Cupboard of the Universe?

I think if I keep trying, I MUST be able to make gold! Just because I am so determined!

ALCHEMIST

The confusion of the alchemists is very understandable. Metals don't come with labels on them telling you what they are made of. Bronze and copper were both metals. Bronze could be melted down into its two ingredients: copper and tin. The alchemists wondered why copper couldn't be boiled down into its ingredients. They didn't know that copper, like gold, is an element. It was only after years of experimenting that ancient scientists began a list of substances that they believed could not be reduced down any further. During the 300 years between 1200 AD and 1500 AD, the list of substances believed to be basic ingredients of the universe grew to include carbon, sulfur, iron, copper, silver, tin, mercury, lead, arsenic, antimony, bismuth, zinc, platinum and gold. (The alchemists eventually had to give up and admit that gold had no recipe.) All of these elements are in our kitchen cupboard of the universe, though they are not sitting all in a row.

Several hundred years went by with no new discoveries of any more basic ingredients. Then, in the 1700s, chemists really began to catch on to the idea of elements, and began intentionally looking for them. They were able to add many more substances to the list of basic elements, including hydrogen, oxygen, nitrogen, magnesium, chlorine, cobalt, nickel, bismuth, platinum and tungsten. In many cases, chemists had not yet been able to produce these substances in a pure form, but they were still pretty sure that these things were basic ingredients, not mixtures. The chemists now began calling these basic ingredients "elements," and by the year 1800, the official list of elements also included phosphorus, fluorine, barium, strontium,

Chemistry labs of the 1700s still looked a lot like the alchemists' labs of the Middle Ages.

molybdenum *(moll-IB-den-um)*, zirconium, chromium, and uranium. These elements were still very mysterious, however, and little was known about them.

The list kept growing during the 1800s, and by the middle of the century there were over 60 elements on the list. By now there was no confusion about what an element was. Scientists understood very clearly that elements were the basic ingredients of the universe. An element could not be boiled down into anything else. They also understood that there was probably a limited number of elements, and once they were all found the fun of discovery would be over. Thus. there ensued a sort of international "scavenger hunt" for new elements, with every chemist dreaming of being one of the lucky winners who would find one of the remaining unknown elements.

Amidst all this frenzy for discovering the remaining elements, a Russian chemist named Dmitri Mendeleyev *(men-dell-AY-ev)* (also spelled Mendeleev) began giving chemistry lectures at St. Petersburg University in 1867. As Mendeleyev studied in order to prepare for his lectures, he began to have the feeling that the world of chemistry was like a huge forest in which you could easily get lost. There were no trails or maps, and there were so many trees! It was all a muddled mess of elements, mixtures, oxides, salts, acids, bases, gases, liquids, crystals, metals, and so much more. The subject of chemistry was confusing to his students, and he could see why. There was no overall structure to this area of science. It was just a massive collection of facts and observations about individual substances. Each scientist had a different way of arranging the substances, and that confused students. Some scientists grouped all the gases together, while others grouped them by color, or listed them from most to least common, or even alphabetically. Was one arrangement better than all the rest? Mendeleyev decided that he would search for some kind of overall pattern that could be applied to chemistry, making it easier for his students to learn.

Mendeleyev began by cutting 63 squares of cardboard, one for each of the elements that were known at that time. On each card he wrote the name of an element and all its characteristics: whether it was solid, liquid, or gas, what color it was, how shiny it was, how much it weighed, and how it reacted to other elements. He then laid out the cards in various ways, trying to find an overall pattern. One evening he was sitting, as usual, in front of his element cards, staring at them and trying to think of some new way to arrange them. He had been working on this puzzle for three days straight, without any sleep. Mendeleyev was exhausted as he fell asleep that night. While he slept, he dreamed about the cards. In his dream he saw the cards line up into rows and columns, creating a rectangular "table."

When he woke up, he realized that his brain had solved the problem while he had slept. The way to arrange the elements was first by weight, then by chemical properties. He began laying out the elements in order of their weight, starting with the lightest, hydrogen.

Then came helium, lithium, berylium, boron, carbon, nitrogen, oxygen, and fluorine. The next element was sodium. Instead of putting it next to fluorine, he put it underneath lithium because it had similar chemical properties to lithium. So the second line began with sodium. Then he began filling in with the elements arranged by weight again: magnesium, aluminum, silicon, phosphorus, sulfur, and chlorine. When he got to the next element, potassium, he decided to start a third line, putting potassium right underneath sodium because they had similar chemical properties. Then it was back to listing them by weight: calcium, titanium, vanadium, chromium... As he laid the cards out in order of their weight, every once in a while, or "periodically," he had to go back and start a new row so that elements that had similar chemical properties would be in the same column. This method of arranging the elements became known as the "Periodic Table" because it is a table (chart) that has patterns that repeat periodically.

This is how the main part of Mendeleyev's chart looked.

lithium	beryllium	boron	carbon	nitrogen	oxygen	fluorine
sodium	magnesium	aluminum	silicon	phosphorus	sulfur	chlorine
potassium	calcium	eka-boron	titanium	vanadium	chromium	manganese
copper	zinc	eka-aluminum	eka-silicon	arsenic	selenium	bromine
rubidium	strontium	yttrium	zirconium	columbium	molybdenum	?
silver	cadmium	indium	tin	antimony	tellurium	iodine
cesium	barium					

Mendeleyev ran into some problems with his Periodic Table. It seemed that there were awkward areas where things did not fit perfectly. He guessed that this was because there were cards missing. His set of 63 cards must be incomplete. Mendeleyev started leaving blank spaces in his chart where he believed there was a missing element. He began to predict what these elements would be like when they were discovered. He even gave them temporary names. The empty space under boron and aluminum he named "eka-boron." ("Eka" means "one more" in the Sanskrit language.) The empty space under carbon, silicon, and titanium was "eka-silicon."

Many chemists of Mendeleyev's day laughed at him for trying to predict the discovery of new elements. They did not believe in his Periodic Table and thought he was a fool for making up all these fictional elements—elements that did not even exist!

In 1875, one of Mendeleyev's predictions came true. A new element was discovered by a French chemist with a long name: Paul Emile Lecoq de Boisdaubran. He had decided to name this new element after his country, France, but using a very old word for France: Gall. He named the element "gallium." Mendeleyev listened to the description of this new element and proudly announced that gallium was, in fact, the missing element he had called eka-aluminum. Mendeleyev had already known what this element would be like. It would be a soft, silvery-blue metal with a very low melting point—so low that this metal might even melt in your hand. This is exactly how Boisdaubran described gallium. Some chemists thought this was just a coincidence and waited to see if any more of Mendeleyev's predictions would come true.

Gallium, from Wikipedia, credit for photo: en:user:foobar

After the discovery of gallium, Mendeleyev became braver about making predictions. He announced that sometime soon a scientist would discover a new element that would be a dark gray metal with a weight that was 72 times heavier than hydrogen, a specific gravity of about 5.5, and having the ability to combine with oxygen to make oxide compounds that are very hard to melt even in a hot fire. Fifteen years after this prediction, a scientist in Germany discovered a new metal that he named "germanium" (after Germany, of course). As you might guess, the characteristics of this new metal were exactly what Mendeleyev had predicted! The scientific world was stunned as they compared Mendeleyev's predictions with the actual experimental results for this new metal—they were almost identical. Germanium was Mendeleyev's "eka-silicon." Mendeleyev was happy to have a real name for "eka-silicon" and gladly replaced it with "germanium."

Germanium looks a lot like gallium. (Photo credit: wikipedia article on germanium.)

Eventually, Mendeleyev and his Periodic Table became famous all over the world. He received gold medals and honorary degrees from universities in other countries, and was invited to join important scientific societies. Sadly, however, his homeland of Russia refused to acknowledge him. When his name was presented to the Russian Academy of Sciences he was rejected. Mendeleyev was unpopular in Russia because he said things the Russian government did not want to hear. He told them they needed to be careful with Russia's supply of crude oil because it was a precious resource and would not last forever. He said that Russia's technology was lagging behind that of other nations and they needed to catch up. Sadly, the government didn't really care about improving the country, and they ingnored Mendeleyev's advice.

After Mendeleyev, chemists continued to discover elements. Every time a new element was discovered it was added to the Periodic Table. The number of elements grew from 63 to over 100. Some adjustments had to be made to Mendeleyev's original table in order to accommodate all the new discoveries. They added a middle section, plus two rows at the bottom.

Mendeleyev's table: The Periodic Table as it looks today:

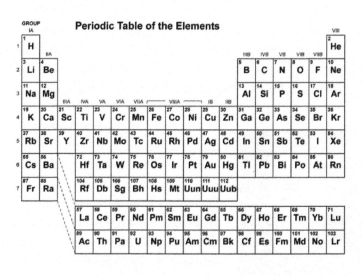

It looks Kind of boring to me.

Yeah, but this is just a black and white one. Wait till you see the bright-colored ones. They're beautiful!

Many decades after Mendeleyev's death, scientists realized that there was nothing on the Periodic Table to commemorate the very man who had created it. So in 1955, when a new element was discovered, the discoverers decided to honor the memory of Dmitri Mendeleyev by naming the new element "Mendelevium." It is number 101 on the Periodic Table and its letter symbol is Md.

To be fair, we really should mention that Mendeleyev wasn't the only person who saw repeating patterns in the elements. A chemist named John Newlands had noticed this in the mid 1800s and published what he called the "Law of Octaves" in 1864 (just a few years before Mendelyev's discovery). Previously, chemists had noticed groups of 3's that behaved similarly and called them "triads." (For example, lithium, sodium and potassium in the first column all reacted violently in water.) Newlands suggested that the triads were part of a larger pattern based on the number 8. He also suggested that atomic weights were a key to organizing the elements. Newlands turned out to be right about both. However, when Newlands presented his theory at the Royal Chemistry Society in London, they laughed at him and even made fun of him. They told him to go play chemistry on a piano.

Unfortunately, this type of thing happens fairly often. New theories that don't fit with current opinions are often scorned or even ridiculed. The Royal Chemistry Society did try to right this wrong in 1884 by asking Newlands to give a lecture at the Society. This time no one laughed at him. And today if you go to the Royal Society of Chemistry website, they proudly suggest that the real discoverer of the periodic arrangement of elements was British, not Russian. In fact, they've even placed a big, blue sign on his birthplace, telling all who pass by that this is where the discoverer of the Periodic Table was born.

Why did Mendeleyev get credit and Newlands did not? Mendeleyev's stroke of genius was to assume that all the elements had not been discovered, and to leave blank spots at points where the pattern seemed to fail. Newlands' table did not leave blanks for undiscovered elements, so it was bound to have problems in the end. Mendeleyev's table was not perfect, either, but it was enough better than Newlands' that Mendeleyev is remembered as the inventor of the Periodic Table.

Activity 2.1 The Periodic Jump Rope Rhyme

Use the first four rows of the Periodic Table as a jump rope rhyme. Why not? Most jump rope rhymes are pretty silly and don't make sense, anyway! The audio track will show you how to say the rhyme. Then try it on your own. If you mess up and trip over the rope, you have to start at the beginning again. Can you get to krypton? Can your friends do it?

Hydrogen, helium,
Lithium, beryllium
Boron, carbon
Nitrogen and Oxygen
Fluorine, neon

Sodium, magnesium
Aluminum and silicon
Phosphorus, sulfur
Chlorine, argon

Potassium, calcium
Scandium, titanium, vanadium, chromium, manganese!

FeCoNi's my CuZn
His last name is Gallium
He lives in Germanium
Once he ate some arsenic, thought it was selenium;
Drank it down with bromine, now he's strong as krypton!

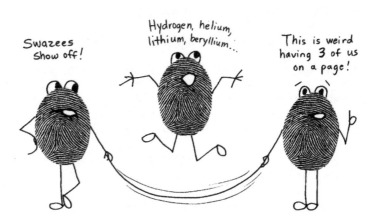

(The audio track can be found at www.ellenjmchenry.com/audio-tracks-for-the-elements, or in the zip file if you bought the digital download version.)

Activity 2.2   What are these elements?

Use the "Quick Six" playing cards to find these elements.

1) Find an element that is used to make matches, fireworks, and detergents. _____

2) Find an element that is used in toothpaste, but is also one of the ingredients in Teflon (The recipe for Teflon is in the "Chemical Compounds Song.") _____

3) Find an element that is found in chalk, plaster, concrete, bones and teeth. _____

4) Find an element that is used in lasers, CD players and cell phones. _____

5) Find an element that is used to repair bones and is also used in paints. _____

6) Find an element that is found in sand, clay, lava, and quartz. _____

7) Find an element that is rose-colored and is used to make catalytic converters and headlight reflectors for cars. _____

8) Find an element that is used as a disinfectant for cuts and scrapes, is used to make lamps and photographic film, and is needed by our thyroid glands. _____

9) Find an element that is used in stadium lights and in large-screen TVs. _____

10) Find an element that is used in dentistry and jewelry, and is also used to purify hydrogen gas and to treat tumors. _____

11) Find an element that is an ingredient of pewter, and can also be mixed with copper to make bronze. _____

12) Find an element that is used to vulcanize rubber and is a component of air pollution. _____

13) Find an element that is used to sterilize swimming pools. _____

14) Find an element that is used in lightbulbs and lasers and won't bond with other elements. _____

15) Find an element that makes up most of the air we breathe. _____

16) Find an element that has no neutrons. _____

17) Find an element that makes diamonds, graphite and coal. _____

18) Find an element that is used in antiseptic eye washes but is also used to make heat-resistant glass, as well as being used in nuclear power plants. _____

19) Find an element that you eat in bananas but can also be used for gunpowder. _____

20) Find an element that is used in lights that need to flash brightly, such as camera flashes and strobe lights. _____

Activity 2.3   Watch some videos

There is a special playlist for this curriculum at the YouTube channel that goes with Ellen McHenry's curricula. The address is: www.YouTube.com/ TheBasementWorkshop.   Click on PLAYLISTS and then on The Elements. There you will find videos for each chapter. The videos have been previewed and pre-selected by Ellen McHenry, so you don't have to worry about inappropriate words or images.   Watch the videos for chapter 2. There are several videos about Mendeleyev, plus a few on gallium and germanium, the elements Mendeleyev predicted correctly.

NOTE: You will notice that there is more than one way to spell Mendeleyev. Some of the videos use the spelling Mendeleev, but it is still pronounced like it has a Y between the two E's. (Recently, spelling it without the Y has become more popular, but it is harder for students to remember how to pronounce it when you don't see the Y.)

Activity 2.4   Alternative Periodic Tables

There isn't any law saying that you can't arrange the chemical elements in some other shape besides a rectangle. Over the past century, quite a few arrangements of the elements have been proposed. We can't print them here due to copyright restrictions, but you can see them online at various websites that have been set up to catalog these alternative Periodic Tables. If you are interested, you can check out these two resources:

1) Wikipedia article: "Alternative Periodic Tables" (pictures are at the bottom)
2) http://www.meta-synthesis.com/webbook/35_pt/pt_database.php

There are also interactive Periodic Tables online, where you can click on an element and information will pop up. Try this one: http://humantouchofchemistry.com/periodic-table.htm

Activity 2.5   Play an online quiz game to help you learn the symbols

You can choose to play easy or harder levels, so this game is great for beginners: http://www.funbrain.com/periodic/

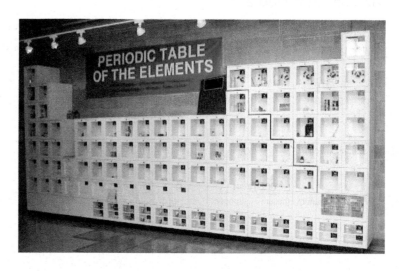

Check out this amazing Periodic Table! It's so large that it covers a whole wall! It is located at the Ruth Patrick Science Education Center in South Carolina. Each box in this table contains an actual sample of the element (except for the elements that are either too dangerous or too rare).

Activity 2.6

Here is a just-for-fun puzzle using the symbols (letter abbreviations) for some of the elements. Write the symbols in the blanks to make some silly riddles.

_____ _____ _____ did the _____ _____ _____ _____a _____
tungsten hydrogen yttrium   molybdenum uranium selenium sulfur yttrium

_____ _____e_____, _____ _____e_____," _____ _____ _____ the
carbon helium phosphorus carbon helium phosphorus tungsten helium nitrogen

_____ _____d'_____ _____ _____e _____ll a_____art? _____ _____ _____
boron iridium sulfur carbon silver iron phosphorus Helium tungsten arsenic

_____ _____l _____ _____g _____ _____ _____r the _____ _____d
fluorine iodine lithium nitrogen indium fluorine oxygen boron iridium

_____ _____ _____ _____ad the da_____ _____ _____ _____.
tungsten hydrogen oxygen hydrogen   yttrium oxygen fluorine fluorine

Here's another one:

_____ _____ _____ do _____ _____ _____ _____t _____
tungsten hydrogen astatine yttrium oxygen uranium germanium tungsten helium nitrogen

_____ _____ _____ _____r _____ _____ _____ a _____ _____ _____ _____e
yttrium oxygen uranium carbon osmium sulfur sulfur   vanadium americium phosphorus iridium

with a _____ _____ _____? A _____rr _____ _____ _____ed _____ _____!
molybdenum uranium selenium   tellurium iodine fluorine iodine carbon astatine

And one last riddle:

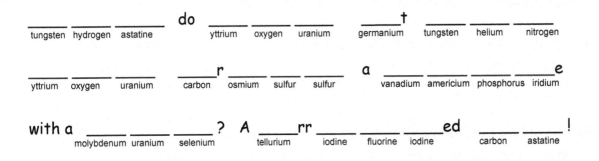

W H at _____ W er e Ba t m a n and R O B i n 's
tungsten hydrogen astatine tungsten erbium Barium thulium nitrogen neodymium oxygen bismuth nitrogen sulfur

ne w N am es af t c r t he y W er e
neon tungsten nitrogen americium einsteinium fluorine tellurium helium yttrium tungsten erbium

ru n O V er B y a C a r?
ruthenium nitrogen oxygen vanadium erbium boron yttrium calcium

F la th a n and R I B B O n !
fluorine lanthanum thulium nitrogen neodymium iodine boron boron oxygen nitrogen

18

CHAPTER 3: ATOMS

The scientists in Mendeleyev's day understood many things about the elements. They had even written books describing the characteristics of certain elements. However, one thing they were never able to do was examine just one particle of an element. It was not until the 1900s that scientists were able to figure out what the elements themselves were made of.

Let's open one of those ingredient jars and find out what the stuff inside looks like. How about He, helium?

ONE PARTICLE (AN ATOM) OF HELIUM:

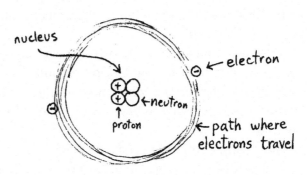

Let's look at just one of those little particles in the jar. One single particle is called an **atom**. An atom is a very strange-looking thing, and is made of up three even smaller types of particles: **protons**, **neutrons**, and **electrons**. The protons and neutrons like to hang out together in the center while the electrons go whizzing around the outside at about a million miles an hour. The central clump is called the **nucleus** of the atom, and the pathways the electrons travel in are called **orbits**, just like the pathways of the planets around the sun.

Your next question might be: "What are these particles made of?" That's a tough question, because they aren't really made of anything—they <u>are</u> the stuff that other stuff is made of. (However, if you asked a physicist this question, he or she would give you a long, complicated answer that included terms such as "up quarks" and "down quarks." Atomic particles such as quarks are still not fully understood and require knowledge of very difficult math. If you want to explore the world of sub-atomic particles, the Internet can help you.) All we need to know is that the proton has a postive electrical charge, the electron has a negative electrical charge, and the neutron has no charge at all. The electron is much, much smaller than either the proton or the neutron. In fact, it is so small that it adds almost nothing to the weight of the atom. When scientists figure out how much an atom weighs, they don't even bother with the electrons. They just count the protons and neutrons.

Let's open another jar. How about Li, lithium?

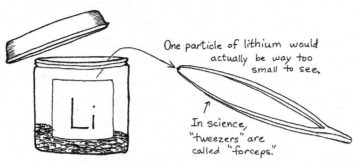

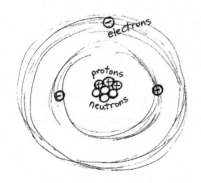

A lithium atom has three protons, three electrons, and four neutrons. The number of neutrons is often the same as the number of protons, but not always, as we see here with lithium. Smaller atoms tend to have equal, or almost equal numbers, but as the atoms get larger they begin requiring more neutrons. By the time you get to number 92, uranium, there are about 50 more neutrons than protons.

You may be wondering why the lithium atom has two rings around it instead of just one, like helium. Why isn't the third electron whizzing around in the first pathway with the other two electrons? The reason is that each ring can only hold a certain number of electrons, and although that first ring may look big enough to you, the electrons think it only has room for two of them. If a third electron comes along, the atom has to add another ring for it. The second ring is a bit bigger, and can hold up to eight electrons.

Let's look at an atom of neon. It has two completely full rings, with two in the first ring and eight in the second ring. Atoms that have full rings with no leftovers are very happy atoms. Neon is very content and good-natured. It is well-behaved and never tries to steal electrons from anyone.

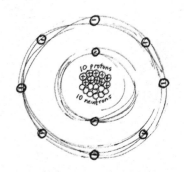

By the way, don't forget that the electrons aren't really as big as they look in these drawings. In fact... now might be a good time to discuss the dimensions of an atom.

If we were to increase the size of an atom until it reached the size of a football stadium, the nucleus would look like a marble sitting on the 50 yard line. The electrons would be smaller than the head of a pin, and they would be whizzing around the outer edge of the stadium.

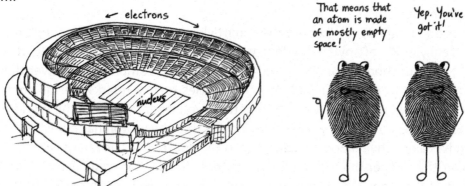

You can see why we have to draw atoms out of scale. If we drew them correctly, either you would not be able to see them, or you would have to have a book about a mile wide.

Speaking of drawing atoms, even if we could make the atoms on this page to scale, they would still be wrong. When scientists finally began to be able to "look" at atoms (though you can't look at them like you can look at a bacteria) they discovered that the electrons don't really go around the nucleus in circles. In the mid 1900s, physicists discovered that electrons move in a more random fashion, not in neat little circles like they had originally imagined. Electrons buzz around so fast that they end up looking more like a balloon than a ring. Since these balloon-like areas looked a bit fuzzy, scientists decided to call them "clouds." Drawing electrons clouds, as we will see, is not easy, which is part of the reason you still see orbitals drawn like rings in most pictures. (The correct name for the "solar system model" is actually the "Bohr model," named after physicist Neils Bohr.)

This is a rough sketch of what electron clouds look like:

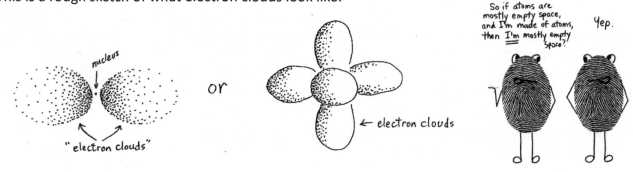

Electrons move around all the time, but they spend more time close to the nucleus than away from it. Electrons like to be in pairs, but in separate clouds on opposite sides of the nucleus. Pairs of electrons hate to be next to each other, too, so each pair will take a position as far apart from the others as possible. The result is that the clouds end up looking like a bunch of balloons tied together, with one electron per balloon.

Large atoms end up having very complicated arrangements of electron clouds and are almost impossible to draw as electron cloud models because so many of the clouds overlap in strange ways.

Activity 3.1

In this activity, you will play the part of an electron. You will map out your location every hour over a weekend. Each hour you will plot your location. For example, if you sleep for eight hours during the first night, you will put eight dots inside the center circle. The next hour might find you in the kitchen eating breakfast, so put a dot there. After that, you might spend an hour watching TV, then three hours at a ball field playing soccer. Put one dot on the TV room, and three dots on the ball field. Continue to plot your locations for several days.

When you are finished, look at your map. Where do the most dots occur? Are they spread out evenly, or is there a definite pattern to the arrangement of the dots? In this model, what represents the nucleus? Do you, the electron, spend more time near the nucleus than away from it?

NOTE: You could just remember a recent weekend (or three weekdays) and estimate the number of hours in various places.

Activity 3.2 Looking at electron cloud models

Look at some pictures of electron clouds on the Internet. Use a search engine with the key words "electron clouds." You will find many different images. Do any of the pictures look like balloons? You may see some that have hourglass shapes or ring shapes, as well as balloon shapes.

Then go to the YouTube channel for this curriculum and watch the videos that show 3-dimensional animations of electron clouds. The atoms will spin around so you can see all sides. Don't worry if there are words or letters that you don't know. Just enjoy watching the animations.

Electron clouds come in four basic arrangements. The first one is spherical in shape and is called an "s" orbital. It seems logical to assume that "s" stands for "spherical," but no... it stands for "sharp." This is very odd, since a sphere is just about the least sharp object a person can think of! The original discoverer of these orbitals was obviously not thinking about their shape when he named them. He was looking at the shape the electron made on something called a spectrograph. However, it's a fortunate coincidence that the word "spherical" also starts with the letter "s," so we can rightly remember the spherical orbitals as being the "s" orbitals.

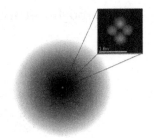

The nucleus is at the center.

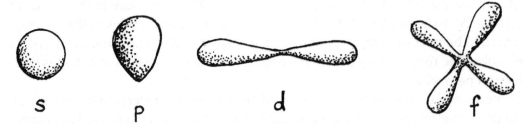

The "p" orbital looks like a balloon. Its name comes from the word "principal" which means "primary" or "the main one." Some people say the "d" orbital looks like a barbell (used in weightlifting) but others say that it looks more like two balloons tied together, albeit very long ones. ("D" stands for "diffuse," another name we'll never remember.) The "f" orbital looks like two "d" orbitals tied together. ("F" stands for "fundamental," which is curious because we already have "principal." It would have been much easier for most of us if they had chosen "a, b, c and d!")

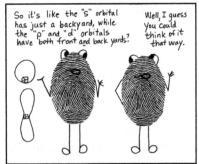

Would you like to see how these orbitals fit themselves together to make "neighborhoods"? The society of electrons has strict rules about how many orbitals can group together to make neighborhoods. The "s" orbital consists of two electrons living together in that one spherical property. (They are the newlyweds who only have eyes for each other and don't mind being alone together.)

The p orbitals group together in bunches of 6. The d orbitals can have up to 10 in their neighborhood, and the f orbitals have a maximum of 14. If we try to draw the neighborhoods, they look something like this:

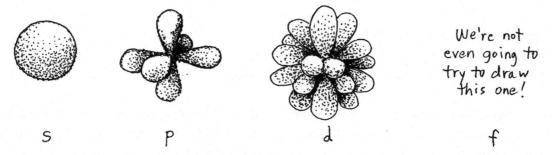

We're not even going to try to draw this one!

s   p   d   f

As you can see, this method of drawing atoms is not easy. If we go back to using the solar system (Bohr) model, things become easier again:

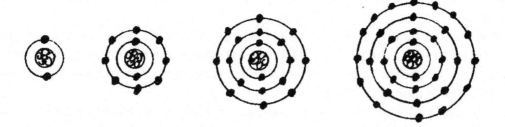

The central clump is the nucleus with the protons and neutrons in it. The dots on the rings are the electrons. Still a bit complicated, but not as bad as the electron clouds. However, this model has its own problems. Most of those rings represent a combination of several electron neighborhoods (orbitals). For example, the second ring out from the nucleus represents an s neighborhood AND a p neighborhood. S orbitals can only hold 2 electrons, and p orbitals hold 6. That second ring has 8 dots on it, so it combines the s and p orbitals into one ring. These rings are often called "shells." The difference between orbitals and shells is often a major source of confusion for chemistry students.

Wouldn't it be nice if there was a way to combine these two methods, the electron clouds and the solar system model? It would clear up all the confusion about orbitals and shells. Fortunately, there IS a way to combine them, and you are just lucky enough to be using the book that invented it!

Activity 3.3   The Quick-and-Easy "Atom-izer"
We are proud to announce one of the most helpful inventions for chemistry students: the Quick-and-Easy "Atom-izer." It lets you "build" atoms so that you can see the electron clouds while using the solar system model. You will find the Atom-izer sheet on the next page, complete with instructions on how to use it. You will need a supply of some kind of tokens that will represent electrons. You could use pennies or pieces of paper, or you might want to add some extra fun by using something edible such as mini-marshmallows or raisins or nuts, so that you can eat each atom you build.

SIDE NOTE: To avoid any scientific confusion, the word "atomizer" is also used to describe a simple device that turns liquids into a spray mist (such as an old-fashioned perfume bottle). We know this, but the name "Atom-izer" just sounded too good to pass up. It sounded like a good name for a device that makes atoms.

Look at the center of the Atom-izer. You will see a dark spot with the letter N on it. This spot will represent the nucleus of our atoms. We know that the nucleus has both protons and neutrons in it, but for this activity we are going to ignore the nucleus. Sorry, nucleus. You'll have to sit out this activity. This is all about electrons and their neighborhoods.

The rules for placing your electron tokens:
1) Always fill smaller rings before larger rings.
2) Always fill "s" orbitals first, before "p" orbitals.

Let's start with the first element: hydrogen. Hydrogen has only one electron. Place a token on one of the black squares in the first ring. It doesn't matter which you choose. Notice that the black squares are not only on the first ring, but they are also on spherical s orbitals. Once you have placed this token, you have made a model of the hydrogen atom.

Now let's make the next element, helium. Helium is number 2 on the Periodic Table and it has 2 electrons. Place a token on the other black square in the first ring. Now you have 2 electrons in the first ring, one on each s.

The next element is number 3 on the Table: lithium. Lithium has three electrons, so you will need to place a third electron token. The first ring is already full, so you will have to go to the second ring. Remember, though, in this second ring you must fill both s spots first. Place a token on either one of the s orbitals. You now have a model of lithium.

Element number 4 is beryllium *(burr-RILL-ee-um)*. It has 4 electrons. Place another token in the other s orbital in the second ring. Presto—you have beryllium!

Boron, element number 5, has 5 electrons. Since the s orbitals in the second ring are now full, you may put your token on one of the black squares on a p orbital. It doesn't matter which p you choose.

Carbon is next. It has 6 electrons, so you will need to add another token to another porbital. It doesn't really matter which you choose, but if you want to be extra-correct, place it in the p that is farthest away from the first p token you placed when you did boron. You see, electrons really don't like to be next to each other unless they have no choice. Given a choice, they will spread out and stay away from each other. So it is best to put the electrons opposite each other.

After carbon comes nitrogen. It has 7 electrons. Add another token to one of the p orbitals. This electron is going to have to be slightly close to another electron. Tough life.

To make number 8, oxygen, add another token to the second ring. To make the electrons as happy as possible, put it opposite electron number 7. Then add a 9th token to make fluorine, and a 10th token to make neon. (By this point, those electrons have no choice but to be next to each other!) Now we have a full second ring. Atoms love to have their rings full. Neon is a lucky atom.

When we go to make sodium, number 11, we will need to put the 11th electron in the third ring. But don't forget—fill those s orbitals first!

Keep adding electron tokens to make magnesium, aluminum, silicon, phosphorus, sulfur, chlorine, and argon.

What if you wanted to go on to element number 19, potassium? You would have to add a fourth ring to this chart. For now, we are going to stop at 18, argon.

Why don't you practice making some atoms "from scratch"? Clear the board, choose any atom from 1 to 18, and build it one electron at time. Then clear it, and try another one.

NOTE: If you need to print out a few extra copies of the Atomizer and you have only a paperback copy, you can download a digital file to print out by going to www.ellenjmchenry.com, clicking on FREE DOWNLOADS, then on CHEMISTRY. You will see a link for "Printable pages for The Elements curriculum."

THE QUICK-AND-EASY "ATOM-IZER"

<u>The rules for placing your electron tokens</u>:

1) Always fill smaller rings before larger rings.
2) Always fill "s" orbitals first, before "p" orbitals.

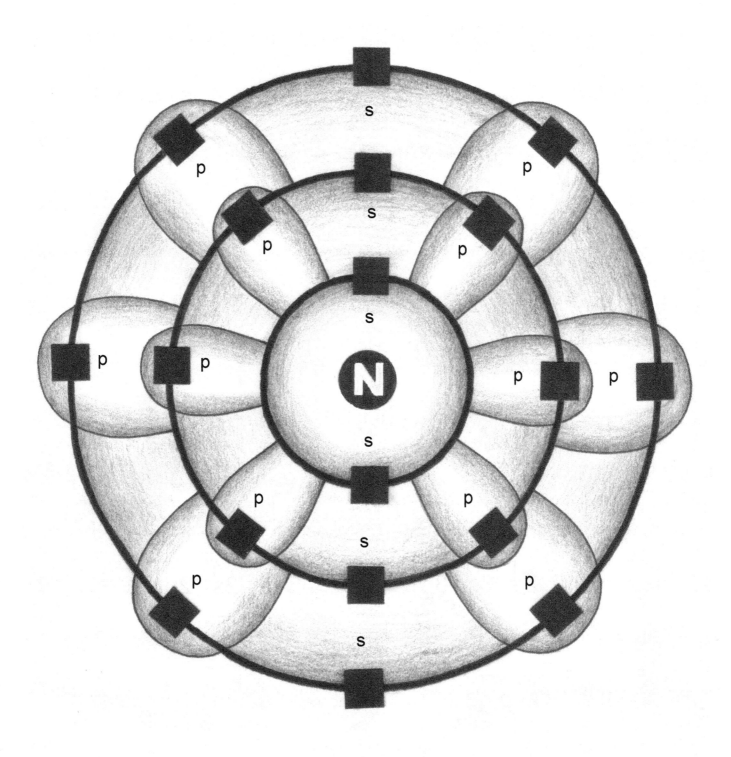

As you can see from the Atom-izer activity, drawing pictures of large atoms would be very time-consuming. So chemists decided to dispense with art altogether and use a string of letters and numbers to show which orbital neighborhoods the atom has, and how many electrons are in each. Their method looks like this: $1s^2 2s^2 2p^6 3s^2$

Look how much space it saves! It's very compact. It means exactly the same thing as a drawing with a whole bunch of rings. Let's look at this method close-up.

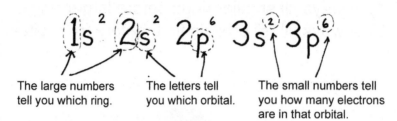

The large numbers tell you which ring.

The letters tell you which orbital.

The small numbers tell you how many electrons are in that orbital.

You can use this method exactly the same way you use the Atom-izer. Instead of the picture with the rings, just fill in the correct number of electrons in each square.

The number of electrons in that orbital goes here.

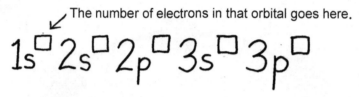

Here is the way chemists would draw oxygen and aluminum:

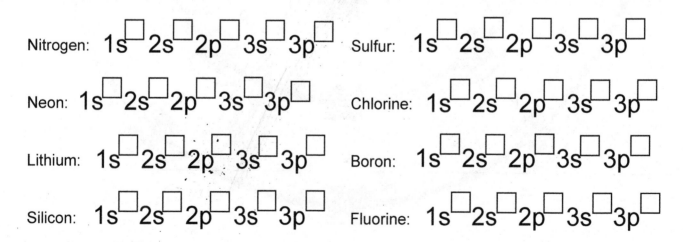

OXYGEN $1s^{[2]} 2s^{[2]} 2p^{[4]} 3s^{[\ ]} 3p^{[\ ]}$

ALUMINUM $1s^{[2]} 2s^{[2]} 2p^{[6]} 3s^{[2]} 3p^{[1]}$

Activity 3.4

Can you switch over from the Atom-izer to this new notation? We've listed some atoms for you to try. All you have to do is write the number of electrons, instead of placing tokens.

Nitrogen: $1s^{\square} 2s^{\square} 2p^{\square} 3s^{\square} 3p^{\square}$ Sulfur: $1s^{\square} 2s^{\square} 2p^{\square} 3s^{\square} 3p^{\square}$

Neon: $1s^{\square} 2s^{\square} 2p^{\square} 3s^{\square} 3p^{\square}$ Chlorine: $1s^{\square} 2s^{\square} 2p^{\square} 3s^{\square} 3p^{\square}$

Lithium: $1s^{\square} 2s^{\square} 2p^{\square} 3s^{\square} 3p^{\square}$ Boron: $1s^{\square} 2s^{\square} 2p^{\square} 3s^{\square} 3p^{\square}$

Silicon: $1s^{\square} 2s^{\square} 2p^{\square} 3s^{\square} 3p^{\square}$ Fluorine: $1s^{\square} 2s^{\square} 2p^{\square} 3s^{\square} 3p^{\square}$

Now for a very cool chemistry fact. The Periodic Table itself can be your guide to electron orbitals. If we were to circle the elements with similar electron orbitals, it would look like this:

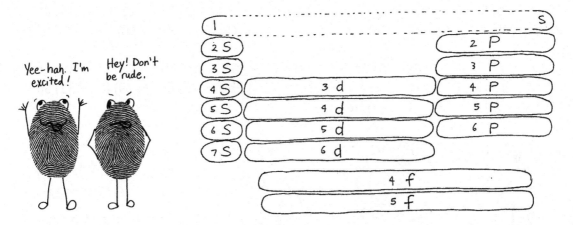

Changing the subject a bit, you may have noticed (during the Atom-izer activity) that the number of electrons seems to correlate with the atom's atomic number on the Periodic Table. This is, indeed, the case, but for a different reason. An atom's atomic number is determined by how many <u>protons</u> it has, not electrons. Each type of atom has a unique number of protons. For example, gold is number 79 on the Table. That means it has 79 protons. Gold is the only type of atom that has 79 protons. If you find an atom that has 79 protons in the nucleus, it's gold. If you added a proton to gold (or took one away) it wouldn't be gold any more. (Such a shame the alchemists didn't know this!)

Since atoms need to be electrically balanced with an equal number of positive and negative charges, they need to have the same number of electrons and protons. This works out nicely for chemists, because they are counting electrons all the time. It's very easy to just look at the Periodic Table for the atomic number and know that not only is it the number of protons, it is also the number of electrons. (Having said this, atoms get out of balance a lot and very often have more or less electrons than protons. We'll see this happen in future chapters.)

Activity 3.5   How many protons?

For each of the following elements, write how may protons it has in its nucleus. (Hint: Remember that the atomic number is defined as the number of protons an element has.)

Ag- silver _____   H- hydrogen _____   Os- osmium _____
Am- americium _____   He- helium _____   P- phosphorus_____
At- astatine _____   I- iodine _____   S- sulfur _____
As- arsenic _____   In-indium _____   Se- selenium _____

Activity 3.6   Guess the atom

This is sort of activity 3.4 in reverse. Can you look at these electron configurations and determine what atoms they are? (Hint: The number of electrons is the same as the number of protons, and the number of protons is the atomic number.)

1) $1s^2\ 2s^2$ _____   2) $1s^2\ 2s^2\ 2p^3$ _____
3) $1s^2\ 2s^2\ 2p^6\ 3s^1$ _____   4) $1s^2\ 2s^2\ 2p^6\ 3s^2\ 3p^4$ _____
5) $1s^2\ 2s^2\ 2p^6\ 3s^2\ 3p^3$ _____   6) $1s^2\ 2s^2\ 2p^6\ 3s^2\ 3p^6\ 4s^2$ _____
CHALLENGE: $1s^2\ 2s^2\ 2p^6\ 3s^2\ 3p^6\ 3d^6\ 4s^2$ _____ (looking at the chart at the top of this page might help)

Activity 3.7 Time to review!

Use the symbol clues to write in the names of the elements.

ACROSS: 3) Co   6) Si   7) U   12) Cu   14) Na   13) Cl   20) Fe   22) O   25) P   27) Mn
         28) I   30) Pu   33) Pt   36) Au   37) Ag   38) Os   39) Hg   40) Ne   41) Li

DOWN: 1) Zn   2) Rn   4) As   5) Ar   8) Ni   9) Kr   10) Mo   11) S   13) Ca   15) Xe
         16) W   17) B   19) H   21) Np   23) N   24) Pb   26) Be   29) F   31) Sn
         32) Mg   33) K   34) Al   35) He

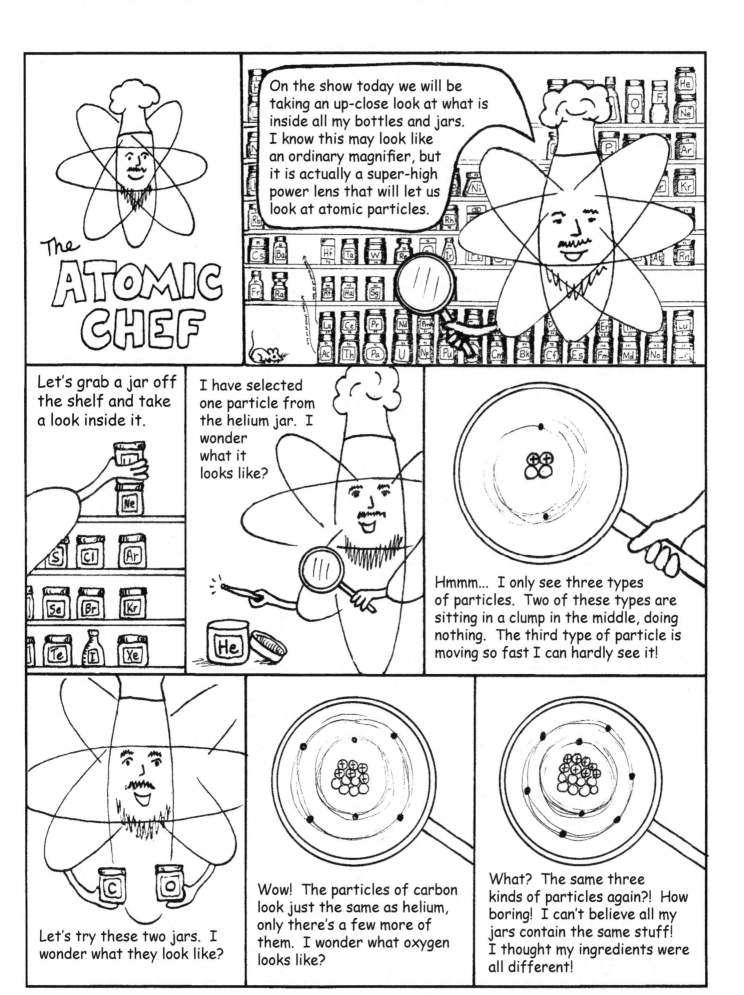

32

CHAPTER 4: MORE ABOUT ATOMS

We will now learn more about the incredibly interesting and wonderfully amazing subject of electrons. The reason this subject is so important is because it's the arrangement of the electrons in the orbitals (especially the ones in the outer ring) that give each element its chemical characteristics. You can suck helium out of a balloon without hurting your lungs because of the way helium's electrons are arranged. Pure chlorine gas is poisonous because of the arrangement of its electrons. If you stick a metal fork in an electrical outlet, you'll get a shock because of the way the electrons are arranged. Carbon can form thousands of different compounds (many of them organic, living molecules) because of the arrangement of its electrons. Understanding the electrons is the key to understanding the chemistry of every substance.

Here are the basic rules that electrons live by:

1) **Spin!**
2) **Always try to pair up with someone of the opposite spin.**
3) **Get plenty of privacy—stay away from other electron couples!**
4) **Try to live in a perfect neighborhood, which is often a group of 8.**

These rules were discovered by combining very complicated mathematics with high-tech scientific experiments. It's kind of funny that the rules sound so simple and yet are based on very complicated math and physics. These four rules are the key to understanding many aspects of basic chemistry.

Chemists often refer to the rings of electrons in those solar system (Bohr) models as "shells." It is the outermost shell (ring) that interacts with the environment around the atom. The electrons in the inner shells just sit there. They almost never come into contact with other atoms. The outer shell is where all the action is.

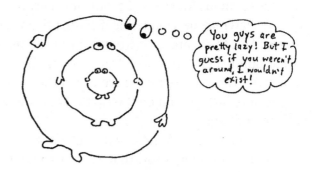

The most important thing to know about outer shells is that the electrons in them take rule #4 very seriously. They are almost neurotic about it. Atoms in the first three rows of the Periodic Table live by the motto: "8 is great!" If there is only one electron in the outer shell, that electron is so miserable that it would rather go off and join another atom than be alone in the outer shell of its own atom. If an outer shell has seven electrons, which is just one short of perfection ("8 is great!"), those seven will try anything to get an eighth electron in the shell. They will even try to steal an electron from the outer shell of any atom that comes close enough. Even when there are two electrons in the outer shell, the electrons are still not very happy. You'd think that they would be content because they have an electron buddy to form a pair with, but those two electrons are still lonely and will look to join with six others to form an "octet." ("8 is great" is often called "the octet rule.")

The number of electrons that an outer shell wants to get (or wants to get rid of) in order to have a full outer shell is called the **valence** number. If the atom needs more electrons, we say that it is minus that number, and we use a minus sign (-). For example, an atom that has seven electrons in its outer shell and only needs one more to make eight has a valence of -1. If it has six in its outer shell, and therefore would like two more, it has a valence of -2.

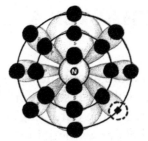

Chlorine
has 1 empty space.

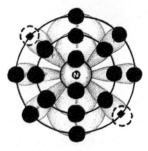

Sulfur
has 2 empty spaces.

If an atom has only one electron in its outer shell, its chances of finding seven more are pretty low. One or two, maybe... but seven? The atom gives up. It's a lot easier just to get rid of that one electron. We would say that this atom has a valence of +1. It has one to give away. Once it gets rid of that electron, the outer shell will then be empty, thus making the next shell down (which is full) the new outer shell. An atom like this can really be obnoxious. It is so desperate to get rid of that one extra electron that it will throw it at any atom that is nearby. (Chemists say "very reactive" instead of "obnoxious.") All of the elements in the first column on the Periodic Table have one extra to give away, so they are all extremely reactive.

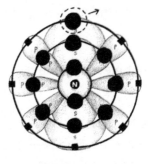

Sodium
has 1 to give away.

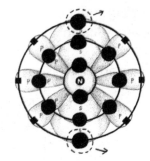

Magnesium
wouldn't mind giving 2.

What about if the outer shell has four electrons? Would the atom try to get four more, or would it give away the four it has? In this case, the atom wouldn't really give or take in the way that sodium and chlorine do. It would be more accurate to say that it forms four bonds, and leave it at that. If an atom has four electrons in its outer shell, we would say that it has a valence of plus or minus four: ± 4. Carbon and silicon are good examples.

You may be wondering if there are any completely happy atoms out there. Yes, there are six incredibly lucky elements on the Periodic Table. Their outer shells are full and they are happy; they are completely non-reactive. These lucky atoms are called the "noble gases." You can find them in the very last column on the right. They are helium, neon, argon, krypton, xenon and radon.

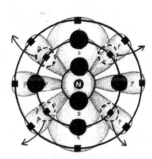

Carbon
has a valence of ± 4

You are very familiar with helium, and you've certainly heard of neon lights. You probably know that Superman is killed by kryptonite, which is a completely fictional substance made up by the cartoonists, and has nothing to do with krypton. Krypton, argon and xenon are used in very bright light bulbs, such as camera flashes. Radon is famous for lurking in basements and causing health problems because of its radioactivity. (Radon's radioactivity isn't an issue with its electrons.) These noble gases have a valence number of 0 because they don't want any electrons and they don't have any to give away. They are perfectly content. Why are they called "noble" gases? They aren't royal, of course. However, if you give human personalities to atoms and say that stealing electrons is bad behavior, then these gases certainly are noble in that they stay out of fights and squabbles over electrons.

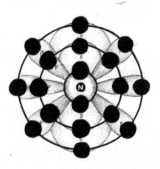

Argon
is perfectly happy.

Activity 4.1   Determining valence numbers

Determine the valence number of these atoms. We've drawn only the outer electron shell because you don't need to see the inner ones to figure out the valence. Remember, the valence is the number of electrons an atom wants to get, or to give away, in order to have 8 in the outer shell. (Remember rule #4: "8 is great!") (+) means extras to get rid of, and (-) means empty spaces to fill. Choose the smaller number as the valence. For example, if you have a choice between a valence of +6 or -2, choose -2 because 2 is less than 6.

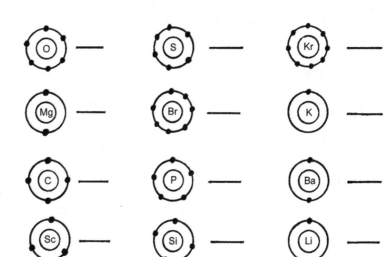

Activity 4.2   Lewis dot diagrams

We don't even need to draw those rings in order to show the valence electrons. (Scientists find all kinds of short-cuts!) All you need to do is write the letter symbol for the atom, then put electron dots around it, with a maximum of two per side (top, bottom, left, right). This is called a Lewis dot diagram.

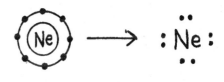

Look at the Lewis dot diagrams and figure out the atom's valence number. This is almost exactly what you did in the previous activity. The only difference is that there are no rings, just electron dots.

B: ____   ·P: ____   ·Ö: ____   H· ____

·S̈: ____   Mg: ____   Äl: ____   Li· ____

:N̈e: ____   ·C̈: ____   ·N̈: ____   :F̈l: ____

Now it's your turn to be the artist and draw the Lewis dot diagrams. Each atom is listed with its valence number. Draw the diagram for each.

Li (lithium) +1 _____   C (carbon) +4   _____

B (boron) +3 _____   Be (beryllium) +2 _____

I (iodine) -1   _____   K (potassium) +1 _____

N (nitrogen) -3 _____   S (sulfur) -2 _____

Activity 4.3 Some very silly element riddles

Here are some totally silly riddles about the names of some of the elements. There is no factual information whatsoever in these riddles!   The point of this activity is just to work on memorizing the letter symbols (and have fun doing it). Fill in the name of the element in the blank.

1) Which element puts you to sleep? ___B___

2)   Which element describes an empty cookie jar? ___Ar___

3) Which element describes what dogs do with bones? ___Ba___

4) Which is the smartest element? ___Es___

5) Which element is "far out"? ___Pu___

6) Which element speaks Spanish, French and German? ___Eu___

7) Which element do you need when your clothes are wrinkled? ___Fe___

8) Which element is Superman's least favorite? ___Kr___

9) Which element has wings on its feet? ___Hg___

10) Which element cheers for the Los Angeles Dodgers? ___Cf___

11)   Which element went to a Clown Convention? ___Si___

12) Which element is found in your wallet? ___Ni___

13) Which element plays an equestrian sport? ___Po___

14) Which element is Dorothy's favorite? (The scarecrow and tin man also like it.) ___Os___

Possible answers:

Es	Ni	B	Eu	Po	Cf	Kr
Ba	Ar	Si	Hg Mercury	Fe	Pu	Os

Activity 4.4   An on-line game   (https://phet.colorado.edu/en/simulation/build-an-atom)

The University of Colorado has a webpage that is very helpful for practicing all those numbers on the Periodic Table. First, there's the atomic number, which is the official number of each element and is the number of protons it has. This is very important to remember. The number of protons determines what element an atom is. Then there's the atomic mass number, which is the combined "weight" of all the protons and neutrons in the nucleus. Protons and neutrons have a mass ("weight") of "1." You can figure out how many neutrons are in an atom by subtracting the number of protons from the mass number. However, you'll notice that the atomic mass numbers are often complicated numbers with decimal points. This is because they are averaging the weights of millions of atoms of that element, and here and there you'll find a few atoms that have one more or one less neutron than most. So these "oddballs" have to be figured into the average weight. For atomic weight, just use the closest whole number. Finally, there are the plus and minus signs which signify ions. Add or subtract electrons and see how the electrical balance signs change. You'll get the hang of it!

Okay, now back to serious business...

Let's look at another pattern on the Periodic Table. Here is the Table with only the valence numbers written in. See if you can find a pattern. (It's pretty obvious.)

																	0
+1	+2											+3	±4	-3	-2	-1	0
+1	+2											+3	±4	-3	-2	-1	0
+1	+2	+3	+4	+5	+6	+7	+3	+3	+3	+2	+2	+3	±4	-3	-2	-1	0
+1	+2	+3	+4	+5	+6	+7	+3	+3	+4	+1	+2	+3	±4	-3	-2	-1	0
+1	+2	+3	+4	+5	+6	+7	+3	+4	+4	+3	+2	+3	±4	+5	-2	-1	0
+1	+2	+3	+4														

*NOTE: Many elements have more than one valency (or "oxidation state"). For example, arsenic can be +3, -3, or +5. We chose -3 in order to emphasize the pattern.

For the most part, the elements in a column have the same valence number. Some elements do, however, have more than one valence, especially the elements in the middle of the table, so we had to simplify the table a bit. Only one valence number was chosen for each element so that the pattern would be more obvious. Chemistry is always like this—things that are generally true, but with lots of exceptions. In this book, we are emphasizing the things that are generally true.

This pattern is more than just an interesting mathematical curiosity. Think back to the story of Dmitri Mendeleyev. Do you remember that Mendeleyev could predict what unknown elements were going to be like before they were discovered? He did not know about valence numbers, but he did know that all the elements in a column were strikingly similar. All would react, or not react, with the same substances. All would have similar electrical properties. Although not identical, they would have similarities in color or texture. Mendeleyev found that if he knew the characteristics of the element at the top of a column, he could predict what the elements below it would be like.

These observations about similarities between elements eventually led chemists to divide up the elements into "family groups." Unfortunately, the names of the families are not anything interesting or easy. They sound like chemistry names. The worst thing is that the first two have very similar names, so it is easy to get them confused.

Alkali Metals
Alkali Earth Metals
Transition Metals
Metals
Semi-metals
Non-metals
Noble Gases
Lanthanide Series
Actinide Series

Notice how many "metals" there are. About 85% of all elements are classified as metals. Sodium doesn't seem like a metal since we most often meet it as a component of table salt. Yet when it is isolated, pure sodium looks like a hard, shiny lump. If you'd like to see what the elements look like in their pure form, use an image search with key words "Theodore Gray Periodic Table."

Activity 4.5   Finding the "families" on the Periodic Table

Use the symbol code shown next to each element to color each square. Color lightly so you can still see the letters and numbers. Or, you may want to just trace around the inside of each square with color. You can choose any colors you want. Fill in the colors in the squares next to the family names, so you know what's what.

1 H •											+ Alkali metals ☐ □ Semi-metals ☐							2 He ○

Symbol code:
- **+** Alkali metals ☐
- **∧** Alkali Earth metals ☐
- Transition metals ☐
- **×** True metals ☐
- **L** Lanthanide series ☐
- **□** Semi-metals ☐
- **•** Non-metals ☐
- **○** Noble gases ☐
- **A** Actinide series ☐
- **H** Halogens ☐

Periodic table:

1 H •																	2 He ○
3 Li +	4 Be ∧										5 B □	6 C •	7 N •	8 O •	9 F H	10 Ne ○
11 Na +	12 Mg ∧										13 Al ×	14 Si □	15 P •	16 S •	17 Cl H	18 Ar ○
19 K +	20 Ca ∧	21 Sc	22 Ti	23 V	24 Cr	25 Mn	26 Fe	27 Co	28 Ni	29 Cu	30 Zn	31 Ga ×	32 Ge □	33 As □	34 Se •	35 Br H	36 Kr ○
37 Rb +	38 Sr ∧	39 Y	40 Zr	41 Nb	42 Mo	43 Tc	44 Ru	45 Rh	46 Pd	47 Ag	48 Cd	49 In ×	50 Sn ×	51 Sb □	52 Te □	53 I H	54 Xe ○
55 Cs +	56 Ba ∧	57 La	72 Hf	73 Ta	74 W	75 Re	76 Os	77 Ir	78 Pt	79 Au	80 Hg	81 Tl ×	82 Pb ×	83 Bi ×	84 Po □	85 At H	86 Rn ○
87 Fr +	88 Ra ∧	89 Ac	104 Rf	105 Db	106 Sg	107 Bh	108 Hs	109 Mt	110 Ds	111 Rg							

58 Ce L	59 Pr L	60 Nd L	61 Pm L	62 Sm L	63 Eu L	64 Gd L	65 Tb L	66 Dy L	67 Ho L	68 Er L	69 Tm L	70 Yb L	71 Lu L
90 Th A	91 Pa A	92 U A	93 Np A	94 Pu A	95 Am A	96 Cm A	97 Bk A	98 Cf A	99 Es A	100 Fm A	101 Md A	102 No A	103 Lr A

You will notice that there are an awful lot of metals on the Periodic Table. Almost everything is a metal of some kind or other. The strangest family name is "Alkali." It is pronounced: "AL-kuh-lie" Also, in a future chapter you'll find out why those two lines on the bottom are sitting down there.

38

The Periodic Kingdom

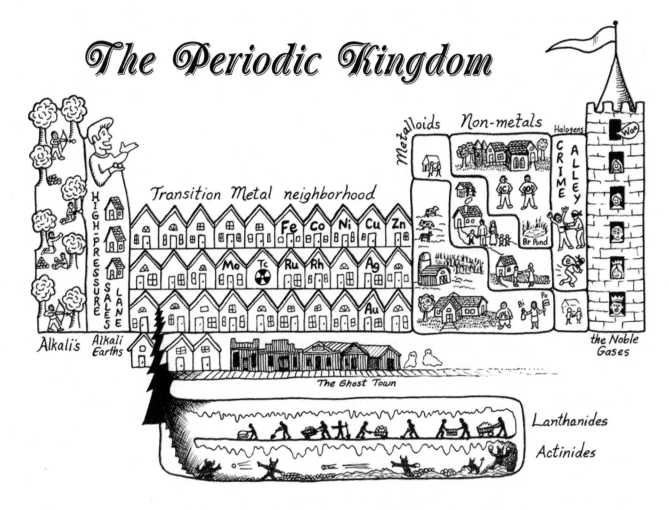

A silly story about real chemistry...

Once upon a time, in a land all around us, was the Periodic Kingdom. The royal family of the kingdom lived on the eastern shore in their tall castle tower. They were the Noble Gases: King Radon, Queen Xenon, Prince Krypton, Prince Argon, Princess Neon, and baby Helium. The Noble Gas family were the most peaceful rulers a kingdom could hope for. They never got upset and never argued with anyone. They remained calm no matter how much turmoil was going on around them.

There was one square mile of land just outside the castle. This land had been divided diagonally, split between two large families. The Metal family set up their homestead in the south western corner. The other family, who lived in the northeast, had such a long last name that no one could remember it, so they became known simply as the Non-metals. Over the years, some members of the Metal family had married members of the Non-metal family. They lived right in the middle, along the diagonal dividing line. This new family was half Metal and half Non-metal, so they became known as the Semi-metals. After several generations, one of the grandchildren decided he did not like being called a name that sounded like "half-breed" so he decided to change the family surname to Metalloid.

In general, the members of the Non-metal family were very conscientious and hard-working. (Without carbon and oxygen, for example, life on earth could not exist.) However, one section of the Non-metal family had gone bad. They all lived on the street nearest to the castle tower. The people of the kingdom called this street "Crime Alley." These poor wretches were always in need. Desperate to gain an electron, they would stop at nothing.

They would even steal or kill to get one. Ashamed to be associated with them, the Non-metals began calling them by a different name: the Halogens. Some folks say this is a sarcastic reference to a halo. Others say the name is based on a statement someone made about them not being worth their salt. Either way, stay out of their neighborhood if you know what's good for you!

Situated in the middle of the kingdom was the town, with all the honest, hard-working laborers. Many of them were miners who earned their living digging for iron, cobalt, nickel, copper and zinc. There were also craftsmen such as silversmiths and goldsmiths. Notable women of the town included Molly, Ruth and Rhoda. Between Molly and Ruth lived a mysterious neighbor who was rarely at home. The strange geometric figure posted on his door made Molly and Ruth nervous, and they told everyone to stay away from it.

On the very western edge of the kingdom lived a band of outlaws named the Alkali Brothers. These outlaws weren't the ordinary type-- they were more like Robin Hood and his band of merry men. They represented generosity gone wrong. The members of the Alkali family have an extra electron they would like to get rid of, but instead of being nice and simply offering it to the poor, they forced the poor to take it whether they wanted it or not. The Alkalis would resort even to violence, if necessary, to get a poor atom to accept an electron! Anyone who came near an Alkali was in danger of being forced to take an electron. (Except the Royals, that is. They lived an enchanted life, unaffected by any of the troubles around them.)

Some of the Alkali outlaws eventually recognized the errors of their ways, tried to reform, and moved a little closer to town. Someone said they had "come back down to Earth" (meaning that they had become more realistic about life) so they added "Earth" to their name and called themselves the Alkali Earth family. However, they still had that Alkali blood running in their veins and that made them pretty pushy. They each had two electrons to give away and they really knew how to pressure folks into taking them!

On the south side of the kingdom was the Ghost Town. The houses all sat there with names on the doors, but mostly there was never anyone home. Once in a while someone would come into town with wild tales about having seen someone in one of the houses for a split second before they disappeared into thin air.

The strangest part of the kingdom was underground. You could only get to it by way of a narrow crack between two of the streets in town. The crack led down into an underground cavern populated by two separate families: the Lanthanides and the Actinides. The Lanthanides were friendly and spent their days mining for rare metals. The Actinides were a treacherous species, and would throw radioactive hand grenades at anyone who came close enough. Beware the Actinides!

CHAPTER 5: MEET THE ALKALI AND HALOGEN FAMILIES

After reading that story, you are probably wanting to know more about these strange families, right? Great! Let's start with the Alkali brothers, those Robin Hood bandits that live on the western shore of the Periodic Kingdom. When we did the Quick and Easy Atomizer, you may have noticed that lithium and sodium both had just one electron in their outer shells:

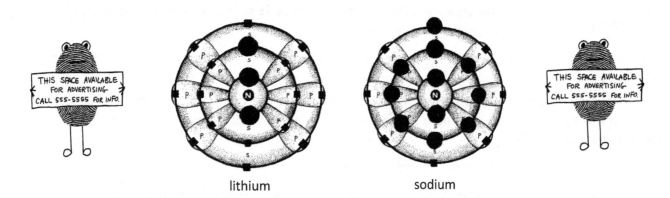

lithium   sodium

Also, you will remember that electrons really don't like to be alone. It goes against one of the rules they live by: "Pair up!" That one electron in the outer ring doesn't think it's worth being part of this atom if it has to be by itself. Therefore, it will try to get away, any chance it gets. It will even try to force itself on atoms that don't want it, which is why these guys are so dangerous. Imagine a group of atoms floating around, minding their own business, when suddenly sodium comes along and "throws" an electron at them. That unwanted electron could cause major disruption. Some substances get very upset when sodium comes near them because sodium can cause quite a reaction—combustion, in fact. If you put a piece of pure sodium into water you will get a spectacular "burning" reaction. Water burning? You might have to see it to believe it.

Activity 5.1

Go to the YouTube playlist and you'll find several vidoes listed under chapter 5 that show what sodium does in water. At least two videos show other alkali metals, not just sodium. The alkali metals underneath sodium are even worse!

Since all the elements in a column tend to behave in the same way (as Mendeleyev knew quite well) we would expect that all the alkali metals would have strong reactions to water. This is true, and, in fact, the reaction gets stronger as you go down the column. Cesium is the most reactive of all. (Francium is radioactive so you just don't play with that one.)

Oddly enough, alkali metals almost never look like metals. If it were not for electrolysis (putting electrical wires into solutions) we would never see pure sodium, and would not know that it could look like a shiny, gray metal. When we meet sodium and potassium in the real world (not in a lab) they don't look like metals at all. They are joined together with atoms of other elements to make molecules such as salt or baking powder. As a general rule, you never, ever, meet the alkali metals by themselves. They are always in the company of other elements. Some elements like gold or silver or sulfur can be on their own, but not the alkali metals.

What about hydrogen? It's in that first column.

On many Periodic Tables, hydrogen is placed right above lithium. Hydrogen has only one electron, so it could be classified as a (+1) valence atom, like the alkali elements in the first column. Hydrogen is super reactive, just like the alkali metals, and is therefore very flammable. (It's famous for exploding and burning in the Hindenburg air ship in 1937.) However, hydrogen is always a gas and therefore can't be solid or shiny like a metal, so it really can't be included in the Alkali family. Some Periodic Tables choose to emphasize the uniqueness of hydrogen, and place it at the top center of the Table, all by itself, instead of in the first column. Putting hydrogen in the center, though, can create artistic problems for graphic designers who are trying to make the table look nice, so H often gets placed above Li, freeing up the center for other things.

Activity 5.2

If you would like to see the Hindenburg burning, there is a video posted in the YouTube playlist. This film was taken by a movie camera that was rolling that day in 1937, intending to record a historic flight, not a disaster. Oops.

The Halogen family, which occupies the column right outside the castle tower, has the opposite problem. They have an outer shell that is missing one electron. They are one short of a full shell and that drives them crazy. So close! They are desparate to get that one last electron to fill that shell, and will try to steal an electron from any atom that comes near. That's why pure chlorine gas is so dangerous. Chlorine atoms all by themselves make a green, poisonous gas. The reason it's poisonous is because of its dire need for an electron. Chlorine would love to find an atom that has an extra electron to give away. Hey, wait a minute—are you thinking what we're thinking? We've got one type of atom that is desperate to get rid of an electron (the alkalies) and one type of atom that is desperate to get an electron (the halogens). The answer is obvious, right? Why not pair these two up?!! Let's introduce sodium to chlorine and see what happens...

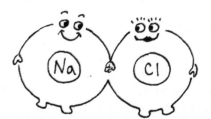

Hey, they like each other! In fact, they seem inseparable—it's a perfect match! The chlorine atom was glad to take the electron that sodium wanted to get rid of.

When two atoms join together, it's called a **molecule**. This is a molecule of NaCl. We know it as ordinary table salt. Remember, we told you that when you meet the alkali metals in real life, they don't look like metals at all. In salt, sodium definitely does not look like a metal, and chlorine doesn't look like a poisonous green gas, either. And though the two of them are dangerous by themselves, when joined together they are safe. You can eat salt and it doesn't hurt you.

The Greek word "halo" means "salt." The word halogen means salt-forming. Any time a halogen connects to an alkali metal, it forms a salt. That would mean that KCl would also be a salt, as well as KBr, LiBr, KI, NaI, CsBr, etc. Any combination of an atom from the alkali metal family with an atom from the halogen family is a salt.

Now this creates some confusion, doesn't it? When chemists talk about "salts" they are not talking about just table salt; they are talking about a whole group of molecules. However, when non-chemists talk about salt, they are usually talking about the stuff we shake onto food. That's why chemists are very careful to say "table salt" when they are talking about NaCl.

There is a special word for this type of bonding, where one atom gets rid of an electron and the other takes it. It's called **ionic bonding**. The root word is "ion." So what's an ion?

An **ion** is an atom that has an unequal number of electrons and protons. We know that an atom all by itself has to be electrically balanced, with an equal number of electrons and protons, so that the positive charges and negative charges are equal. If an atom has either more protons than electrons, or more electrons than protons, it is called an "ion." The word "ion" comes from Greek and means "going." Michael Faraday, one of the scientists who discovered electricity in the early 1800s, chose this name for these unbalanced atoms. Where are the ions going? Faraday saw them going to metal electrodes that were stuck into chemical solutions. Some ions would go over to the postive electrodes and others would go to the negative electrodes.

Let's look at the sodium and chlorine atoms while they are bonded to each other.

Here is an electron/proton count:

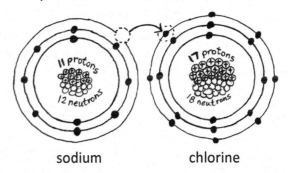

sodium   chlorine

	electrons	protons
Na (sodium)	10	11
Cl (chlorine)	18	17

Sodium has 10 negative electrons and 11 positive protons. Since negative and positive sort of cancel each other (like positive and negative numbers in math), this means that sodium has an overall electrical charge of +1. Conversely, the overall charge of chlorine is -1 because it has one more negative electron than it does positive protons.

Chemists write ions like this: Na^+ or Ca^{2+} They put the overall electrical charge in very small type (called superscript) at the top right of the letters. If the number happens to be a 1, they often leave out the 1 and just put a positive or negative sign, as though the 1 was invisible.

Since our sodium and chlorine ions now have opposite electrical charges, Na^+ and Cl^-, they are attracted to each other and the ions stick together. (Always remember: **opposites attract.**)

What makes table salt safe to eat, when sodium and chlorine are so dangerous? When sodium and chlorine are bonded together, they are both very content. Sodium no longer wants to get rid of an electron, and chlorine has the extra it wanted. Everyone is happy. Therefore, a molecule of NaCl is very safe. You can hold salt in your hand and it just sits there. If you put salt into water, though, you can get the two to separate. Once separated, the two atoms have the potential to be very dangerous again. Water, however, has a way to deal with this. Fortunately, our bodies are full of water molecules, so we can safely eat salt. (In fact, sodium and chlorine ions are essential to our health!)

Activity 5.3   Tear apart salt molecules and put them back together again

For this activity you will need a hair dryer, salt, water, a bowl and two plates. If you have a magnifying lens, add this to your list, so you can get a close-up look at the crystals.

How would you go about tearing sodium away from chlorine? Use microscopic pliers? Actually, all you need to do is put them into a glass of water.   The water molecules will pull the sodium atoms away from the chlorine atoms. We call this process "dissolving."

Put some salt into the water and stir it. The salt will seem to disappear. As the sodium and chlorine molecules detach from each other, the water molecules form "cages" around them.

A water molecule looks like this.

The side with the oxygen atom has a slight negative charge and the side with the hydrogens has a slight positive charge.

The sodium ions in the lattice (the square pattern) have a positive charge. Therefore they are attracted to the oxygen atoms in the water molecules. The chlorine ions have a slight negative charge and are attracted to the hydrogens in the water molecules. Remember, sodium gave away its extra electron. It doesn't get it back in this situation. Sodium goes on without the electron, continuing on as a positive ion. Water is able to coax sodium away from chlorine without it getting its electron back again.

This is what salt water looks like on the atomic level:

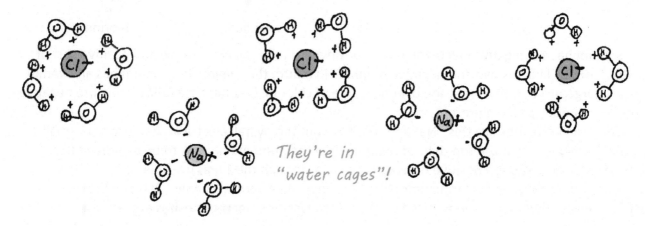

They're in "water cages"!

Pour two puddles of salt water (about an inch in diameter), one on each plate. Make sure they are far enough apart so that you can dry one with the hair dryer without the other one being affected. Leave one puddle to evaporate on its own. Blow the other puddle with the dryer so that it evaporates quickly. What you are doing is removing water molecules.   With the water molecules gone, the sodium and chlorine atoms once again bond together. You should see crystals forming. Observe both puddles after they are completely dry. Which crystals look more like salt crystals? The ones that dried more slowly had more time to make nice crystals. The ones that were hurried with the hair dryer had to do a rush job. (Have you ever had to do something in a big hurry and felt like you could have done it a lot better if you had had more time?)

****

So the alkali metals love to pair up with the halogens. They give and take their electrons and are very happy together. And when they get together, the result is always a salt.

What about the "cousins" of the Alkali brothers? The ones that reformed and decided to be just pushy instead of dangerous? In our story we said that they "came back to Earth" in their attitude, so they became the Alkali Earth Metals. This isn't really the reason the word "earth" is in their name, but the story is a great way to remember their name, and it is true that they are not quite as bad as the alkali metals, but are still very insistent on giving away the two electrons in their outer shell.

The most famous alkali earth metals are magnesium and calcium. How many times have you been told to drink milk because it has calcium in it? (Spinach has calcium, too!) Our bodies need both calcium and magnesium in order to work properly. Magnesium was one of the atoms you made when you did the Quick and Easy Atom-izer. Here it is, on the right, showing its two lonely electons in the outer shell. To fill the shell, it would have to find another six electrons. It's easier for magnesium just to give away those two.

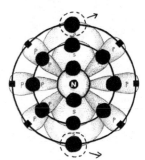

Magnesium

Now let's think about the Periodic Table. What column on the table has atoms that would like to get two electrons? The column to the left of the halogens. This column is made of oxygen, sulfur, selenium, tellurium, and polonium. Let's try matching up magnesium with oxygen and see what happens.

The recipe for what we have made is MgO. The name for it is **magnesium oxide**. You can find this substance occurring in nature as the mineral **periclase**. This mineral can be used in a lot of ways. It is the primary ingredient in some stomach medicines that relieve heartburn. It is used in making fireproof construction materials and as insulation against electrical fires in cables. Both humans and animals can take it as a way to get magnesium in their diet. Magnesium is necessary for your body to heal itself. A magnesium compound called epsom salt can be added to warm water to make a solution to soak in. The magnesium help both skin and muscles.

A sample of periclase

Once again, we have **ionic bonding** going on. Whenever you have elements on the left side of the Table pairing up with elements on the far right side, you get an ionic bond (two atoms giving and taking electrons), and the result is a salt. Would magnesium be just as happy with sulfur as it is with oxygen? Yes, it would. Sulfur also needs two electrons. If you pair magnesium and sulfur you get MgS, **magnesium sulfide**. This substance doesn't have nearly as many uses as MgO. However, if you need to get rid of sulfur, magnesium is always willing to take it. During steel production, iron is heated up so hot that it turns to a liquid. Unfortunately, the molten iron has some other things mixed in with it, and sulfur is often one of them. You don't want sulfur in your iron. How do you strain out the sulfur? One way is to dump powdered magnesium into the molten iron. The magnesium will go around pulling all the sulfur atoms out of the steel. Then the magnesium and sulfur bond together and form MgS, which floats on the surface and can be raked off. Isn't that a clever use of chemistry?

Epsom salt is a combination of magnesium ions (Mg^{2+}) and sulfur ions (S^{2-}), with four oxygens thrown into the deal, too. The recipe is $MgSO_4$. Epsom salts were first discovered in Epsom, England, in the 1600s. However, it wasn't until the 1800s that magnesium was officially discovered as an element and given the name "magnesium."

Sir Humphry Davy was the scientist who discovered magnesium. He also discovered a number of other alkali elements including sodium, potassium, calcium, barium and strontium. He used electricity to pull alkali atoms away from their molecules. Davy used a simple battery (a Voltaic pile) to make positive and negative electrodes that could be put into chemical solutions containing alkali elements. The element ions would go <u>over</u> (remember, "ion" means "going") to one of the electrodes and stick there. In this way a whole clump of pure element could be produced. This is the way the alkali elements were isolated so they could be studied and named.

Davy was quite a showman, too. People would buy tickets to hear him lecture and to see him perform his chemistry demonstrations. One of his most famous demonstrations involved not alkali elements, but nitrogen and oxygen. When these two elements are joined to make N_2O, the result is a gas called nitrous oxide, or "laughing gas."

Alkali earth metals can be a lot of fun when they are added to explosives and shot up into the air. (Fireworks!) Magnesium burns white, calcium burns orange, strontium burns red, and barium burns green. Which one of these alkali earth metals is safe enough to use in sparklers? Is it on the top or bottom of the table? When the alkali metals burn in water, which one is the least dangerous? Where is it located on the table, top or bottom? Do you see a pattern?

The alkali metals also produce colors when they burn. Sodium, for example, burns with a yellow color. Sunlight is a yellow color because there is a lot of sodium in the sun. Sodium is also used in outdoor lighting. Sodium lights burn with an eerie yellow glow. Lithium burns pinkish-red and potassium burns lilac purple.

Activity 5.4   Watch some alkali elements go up in flames

Go to the YouTube playlist and look for the video(s) showing "flame tests." Each element burns with a different color. You can identify the elements in a compound by looking at the color of the flame.

A sample of barite in the form of a "desert rose."

Barium often bonds with sulfur to form **barium sulfate**, $BaSO_4$, also known as the mineral **barite**. (In the UK, it is spelled "baryte.") The Greek word "barus" means "heavy," and barium is, indeed, one of the heavier elements. The heaviness of barium is a great help to the oil and gas drilling industry, for barium powder is added to the drilling fluids in order to prevent blow-outs in wells. Barite is a natural, non-toxic substance, so it does not add any pollution to the environment.

Barium is also very important to the health care industry because it is used to make x-ray pictures of the stomach and intestines. Since barium shows up very clearly on x-rays, doctors will ask a patient to drink a barium liquid before the x-ray is taken. The barium makes every twist and turn of the digestive system visible. The barium solution is not toxic, so no harm is done to the patient.

Meanwhile, back in the laboratory of the Atomic Chef...

CHAPTER 6: THE NOBLE GASES AND THE NON-METALS

Think back to our Periodic Kingdom fairy tale. Do you remember the description of the Noble Gas family? They were the most peaceful rulers a kingdom could hope for. Nothing ever upset them. The real science behind this part of the story is that the noble gases are the only elements on the Table that have the exact number of electrons they want in their outer shells.   Of course, using words like "peaceful" and "happy" to describe something that isn't alive is a little silly, but it does help us to remember the real science.

A chemist would say that the noble gases are "inert," which means they don't react with anything.   An atom of helium doesn't want to give or get any electrons because its outer shell is full. Therefore, it will not interact with the atoms around it. Because it is inert, helium won't react with the atoms in your body, which is why it isn't dangerous to take a breath of it in order to talk funny.

A scientist of the 1800s using a spectrometer

Helium was named after the Greek god of the sun, "Helios," because the sun was the first place that helium was discovered. The discovery was made in the year 1868 using a machine called a **spectroscope**. If you look at a light source through a spectroscope you will see colored lines. Each element has a unique pattern of lines. Sodium's pattern is very simple and consists of basically two yellow lines. (To see the pattern you have to heat or burn the sodium so that it makes light.) Looking at the sun is a bit risky, as it can cause eye damage. The discoverers of helium looked at the sun during an eclipse, when the light was reduced and therefore less dangerous to look at. They saw sodium's

Helium's spectral pattern

two yellow lines, plus many others that they recognized, but they also saw a new pattern they didn't recognize. They understood that what they were seeing was probably a new element and they immediately named it helium, thinking it was a special element found only in the sun. Then, in 1903, large amounts of helium were found mixed in with natural gas deposits. Later, it was also found mixed in with uranium ore. Helium was definitely part of the earth's natural chemistry, not just the sun's. (Later, it was discovered that helium is a by-product of radioactive decay.)

Attribution: Atlant, on Wikipedia "xenon"

Because the noble gases are inert, they are ideal for use in light bulbs. They will not ignite or explode and are safe even when exposed to electrical currents. Neon is (obviously) used in neon lights, argon is used in ordinary bulbs, krypton is found in fluorescent bulbs and camera flashes, and xenon is put into ultraviolet lamps, camera flashes, and lighthouse bulbs. The xenon bulb shown here is used to project IMAX films, which require a very bright light source. Helium, neon and argon are also found in the most common types of lasers.

You may have heard that radon causes lung cancer. This noble gas is inert, just like all the others, so why does it cause problems? The problem with radon is not its electrons, but its nucleus. Radon is radioactive, which means its nucleus is throwing out harmful particles. Radon occurs naturally in the earth, especially in areas that have lots of uranium. Mostly, radon just goes up into the air and gets lost in the atmosphere. It only causes problems when a whole lot of it seeps up into the basement of a building or into a mine shaft.

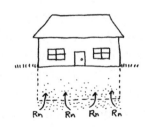

We already talked about one section of non-metals, the halogens. The other members of the non-metal group are carbon, nitrogen, oxygen, phosphorus, sulfur, and selenium. Some

chemists also include boron in the non-metal group. We could also include hydrogen as a non-metal if we wanted to. Hydrogen is sort of a group unto itself, but it is found connected to carbon and oxygen so often that we could legitimately think of it as a non-metal.

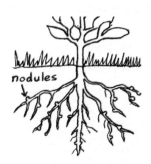

Nitrogen makes up almost 80% of the air we breathe. Nitrogen gas is made of two atoms of nitrogen bonded to each other to form N_2. The nitrogen we breathe doesn't do anything but take up space in our lungs. It's the oxygen that we need from the air. We do need nitrogen in our bodies, though, as it is an essential ingredient in proteins. The nitrogen atoms found in proteins come from the food we eat, however, not the air we breathe. Plants have a similar situation. They need nitrogen to make chlorophyll, but they can't get the nitrogen out of the air. Nitrogen is all around them, but the plants can't use it. For a plant to be able to take in nitrogen, the nitrogen atoms must be attached to molecules in the soil. Fortunately, there are bacteria in the soil that are capable of taking nitrogen out of the air and putting it into a form that plants can use. They are called "nitrogen-fixing" bacteria. You can find these bacteria growing in colonies on the roots of certain plants. The bacteria colony will look like a little bump about the size of the head of a pin. These bacteria prefer to grow on the roots of beans, peas, peanuts, soybeans and clover. Ancient farmers knew that these plants enriched the soil, but they did not know why. They rotated their crops each year, so that each field would get a turn having a bean crop in it. The beans would restore the nitrogen to the soil. In modern times, farmers just put nitrogen fertilizer on their fields.

If you have a super-powerful refrigeration unit, and can cool nitrogen down to several hundred degrees below zero, it turns into a liquid—a very cold liquid, so cold that it can freeze things instantly. As soon as it is exposed to air or water, it boils and evaporates, returning to its gaseous state, returning to the air from whence it came.

Activity 6.1 Fun with liquid nitrogen

Liquid nitrogen isn't easy to get. It takes a special (expensive) refrigeration unit to get the temperature down to hundreds of degrees below zero. Fortunately, some folks who do have access to liquid nitrogen have filmed their demonstrations and posted them on the Internet. Go to YouTube.com/TheBasementWorkshop and you'll find some liquid nitrogen experiments on The Elements playlist.

Oxygen makes up about 20% of the air we breathe. Just like nitrogen, oxygen goes around in pairs (O_2). A single oxygen atom is a very unhappy atom because it has two empty electron slots in its outer shell. One oxygen atom by itself is very dangerous. We've learned to use this to our advantage, though, when we want to get rid of germs. Bleach, NaClO, has that single oxygen atom hanging on the end of the NaCl, and it can fall off very easily. When the oxygen atom falls off, it goes about looking for electrons. It will steal electrons from anything nearby, hopefully a germ that we want to kill anyway. A whole bunch of single oxygens can wreck a bacteria's molecules so badly that it dies.

"Help! I'm surrounded by single oxygens!"

When two oxygen atoms pair up as O_2, the electron math doesn't work out perfectly. If each oxygen atom wants to get 2 electrons, then how can they be happy together? They work out an arrangement where they each share one of their electron pairs. Electrons move so fast that they can almost be in two places at the same time. Almost. So for a split second, one oxygen will have its own 6 electrons plus the 2 it is borrowing, to make 8 in the outer shell. For that split second it is happy. Then it must return the favor and share a pair with the other atom. This would mean that for a split second it would only have 4. But before it can get really unhappy about that, it's suddenly time to receive again, and it finds itself with 8 for another split second. This back and forth sharing happens so fast that the atoms feel like they have 8. Or at least they feel like they have 8 just often enough to prevent them from splitting up into singles. However, the fact that they are not completely content with the situation is also what makes them so useful to living things. They can be split up and used for many biological processes. Oxygen is also necessary for the energy-releasing process of combustion (burning).

Oxygen is a main ingredient in many minerals. Oxygen bonds with silicon to make silicon dioxide, SiO_2. Sand, glass and quartz crystals are made of SiO_2. The minerals hematite (Fe_2O_3) and magnetite (Fe_3O_4) are commonly found in the earth's crust. Even ice, which is frozen H_2O, can be classified as an oxide mineral.

A hematite carving

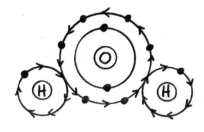

Speaking of H_2O, let's take a look at how these atoms stay together. We won't see an ionic bond here. Ionic bonds are formed only by the elements on the far sides of the table. An atom on the left side pairs up with an atom on the far right side, such as sodium and chlorine. Oxygen does not make ionic bonds. Non-metal atoms form a type of bond called covalent. In **covalent bonds**, electrons are actually shared, not given away. We won't see any electrically unbalanced atoms here. The atoms match themselves up so that they can all share their electrons. For example, oxygen has 6 electrons in its outer shell and hydrogen has 1. Two hydrogens can get together with one oxygen and all three of them together have a total of 8 electrons. The 8 electrons circulate around (at lightning speed) and make sure all the atoms are happy. (Of course, hydrogen is very small and doesn't want 8. It only wants 2.)

Another example of covalent bonding is carbon dioxide. One carbon atom gets together with two oxygen atoms. The oxygens would like to gain 2, and the carbon doesn't mind sharing its 4. As electrons move very quickly, the atoms can manage to share the 8's.

Sulfur also has 4 electrons in its outer shell. Sulfur can bond with two oxygens, just like carbon can. SO_2 is called sulfur dioxide. (You may be catching on by now that "di" means "two.") Sulfur dioxide

AND THE WINNER IS...
VOLCANO.!

is a poisonous gas. When it is released into the air (often by coal-burning factories), it causes air pollution and acid rain. It's not just humans that make sulfur dioxide, though. Volcanoes make far more of it than any factory does.

Now here is a very strange covalent molecule: H_2O_2, hydrogen peroxide. This is the stuff that looks like water but is used for first aid, to clean cuts and scrapes on your skin. Let's do an electron count. The hydrogens each have 1 electron, and the oxygens each have 6. That's 6+6+1+1=14. Whoa! How can that be?

And we were so happy as H_2O...

Hydrogen peroxide is water with an extra oxygen stuck on. Water is perfectly content the way it is. Why would it want another oxygen stuck onto it? This is another case of an oxygen atom that can easily fall off its molecule and become a dangerous single oxygen. If you want to kill germs, single oxygens can really help!

Phosphorus was first discovered in the year 1669 when a chemist was boiling a batch of... urine. No kidding, he collected hundreds of gallons of pee and was going to boil it until it turned into gold. (Well, urine is yellow, gold is yellow—could be a connection there.) What his experiment produced was far more amazing than gold. It looked like a disgusting lump of yuck (and it smelled terrible) but when he heated it, it glowed with a brilliant white light. Back in the 1600s they had never seen a light bulb, so glowing phosphorus must have seemed almost magical. He had discovered one of phophorus' more interesting qualities. The name phosphorus means "light-bearer."

White phosphorus in a jar of water

Pure phosphorus can be either white or red. In white phosphorus you find 4 atoms binding together to cope with their three empty electron slots. Eventually, white phosphorus turns into red phosphorus as those foursomes split apart. You've seen red phosphorus on the tips of matches. Match heads also contain sulfur, another non-metal.

Phosphorus is also involved in energetic tasks in living cells. When combined with oxygen, it's the P in the ATP—a molecule that acts like a rechargeable battery. Phosphorus is necessary in other biological process, also, so it is an element essential to life.

Carbon is the most amazing atom in the non-metal group. Because it has a valence of +4 or -4, it can bond with itself or with other atoms in all kinds of ways. When carbon bonds with itself, it can make something as humble and inexpensive as graphite (the "lead" in pencils) or as valuable as a diamond. It may be hard to believe, but graphite and diamonds both have the same chemical recipe: just carbon. How, then, can they be so different?

To discover the answer we must look at how the carbons are bonded to each other. In the case of diamond, the basic geometrical shape looks like a pyramid. When millions upon millions of these molecules are bonded together like this, we get a diamond. The bonds in this shape are very strong, which is what makes diamonds so hard.

Another shape that carbon can bond into is a six-sided hexagon. Graphite is layer upon layer of flat sheets of connected hexagons. The sheets are only loosely held together, and can slide back and forth. This is why pencils rub off onto paper, and why graphite can be used as a dry lubricant. (Graphite from a pencil can be rubbed onto the bottom of wooden dresser drawers to make them slide in and out more easily.) Technically, if you could squeeze the graphite in your pencil hard enough to make the carbons change their geometry from hexagons into pyramids, you could make a diamond.

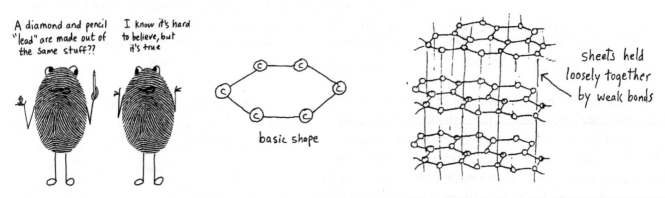
A diamond and pencil "lead" are made out of the same stuff??
I know it's hard to believe, but it's true
basic shape
sheets held loosely together by weak bonds

The most fantastic shape carbon can make looks exactly like a soccer ball. Sixty carbons can join together to form a sphere made of hexagons and pentagons. Since this shape looks a bit like the geodesic domes used in architecture, it was named after an architect famous for designing dome structures, Buckminster Fuller. Chemists decided to name this molecule "buckminsterfullerene," or "buckyball" for short.

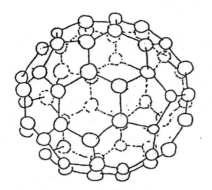 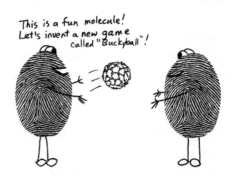

Carbon is the central atom in all organic molecules. We call molecules "organic" if they are based on carbon. Some types of organic molecules are found in plants, animals and microorganisms. It is the "backbone" of DNA, proteins, sugars and starches. Other kinds of organic molecules aren't found in living things. The molecules that plastic is made of, for example, are called organic because they contain long chains of carbon atoms. Gasoline and other petroleum products are also carbon-based and are therefore classified as organic. (If you'd like to learn more about all the amazing things carbon can do, there's a sequel to this book called *Carbon Chemistry*.)

Pure sulfur is a yellow solid.

Sulfur is right under oxygen on the Periodic Table. This means it also has 6 electrons in its outer shell, and would like to gain 2 more to make 8. It will, therefore, have some chemical similarities to oxygen. However, sulfur atoms are larger than oxygen atoms, having an atomic mass (weight) double that of oxygen. Larger atoms are less likely to be gases. (Krypton, xenon and radon are curious exceptions to this rule.) Pure sulfur is found as a yellow solid and has a strong odor. That's one of sulfur's characteristics—it smells. Sulfur is found in many organic molecules in both plants and animals. It's the key element in the stink of skunks and garlic. When eggs go rotten, they produce hydrogen sulfide, H_2S, which smells bad in a sulfur-ish way.

Garlic gets it smell from the sulfur compounds it contains.

Sulfur is part of several amino acids (the stuff that proteins are made of). Its presence in hair proteins makes hair waterproof. Sulfur can allow molecules to make "cross-bridges" which makes them tougher. It can be added to rubber to keep it from melting in high temperatures and cracking in low temperatures.

Selenium is right under sulfur on the table, which means it has the same number of electrons in its outer shell. Since selenium has the same valency as sulfur, it is sometimes found in minerals that usually contain sulfur, with selenium taking the place of sulfur. The metal atoms in these minerals are happy with either sulfur or selenium; it doesn't make a big difference to them. Both S and Se want to make 2 bonds, and that's the most important issue to the metal atoms. They'll bond with either one.

Selenium's name comes from the Greek word for the moon, "selene." Selenium doesn't have any features of the moon. It seems that the discoverer of selenium noticed its striking similarities to the element tellurium, right underneath it on the Table. He thought that since tellurium was named after the earth, perhaps the "earth element" should have a "moon element" nearby. Or so the story goes.

Selenium is named after the moon.

Selenium is found in some key molecules in our bodies, but it is not as abundant as oxygen and sulfur. Some people take selenium supplements because selenium is said to be able to help clean up "free radicals" in the body. Free radicals are dangerous fragments of molecules. Selenium used to be used quite a bit in the electronics industry, but now silicon has taken over. Selenium is still used by the glass making industry, and it is also a key ingredient in solar cells.

The halogens (fluorine, chlorine, bromine, iodine and astatine) are a subset of the non-metal group. We can think of them as non-metals, or as the salt-making halogens. Both are correct.

Some chemists like to put boron, silicon, arsenic and tellurium into the non-metal group, as well. This causes a lot of confusion for chemistry students. If you do a search on the Internet for Periodic Tables, you will find that some tables color code these elements to be in the non-metal group. Other tables will have them color coded to match the metal group along with aluminum and tin. And still others will split the difference and put them into their own group called the semi-metals. Who should we believe? In the end, it doesn't matter too much how they are classified because the elements don't care, and classification doesn't change them in any way. They are what they are, no matter what we call them. Perhaps the most important lesson to learn here is that scientists don't always agree!

Activity 6.2   Watch a demonstration that uses noble gases

Except for helium, noble gases are not items we can just pick up at the store. Therefore, we have to rely on generous scientists who take the time to post their noble gas demonstrations on YouTube. There should be one or more posted for you at the Elements playlist. One of them shows six balloons, each one filled with a different noble gas. What will happen when the demonstrator lets them go? Helium is easy to predict, but what about xenon?

If you like silly animated cartoons, there is also a cartoon music video made by a student (perhaps not too much older than yourself) with funny rhymes and pictures about the noble gases.

Activity 6.3   A famous silly song about the elements

A number of years ago, an entertainer named Tom Lehrer wrote and performed "The Elements Song." The lyrics of the song are simply the names of the elements, rearranged so that they rhyme. (This means they are not in order, so you can't use this song to memorize the Table.) There are several versions of this song posted on The Elements playlist on the YouTube channel. One version is a historical film (in black and white) of Mr. Lehrer performing his song for an audience. Another version provides a nice picture of each element as it is named, and a third version has the song slowed down so you have a better chance of being able to sing along.

Activity 6.4   A puzzle about carbon-based molecules

There's nothing like carbon when you want to form bonds. Carbon bonds in more ways and with more elements than anything else on the Periodic Table. It's the "nice guy" among the elements. You could imagine it being willing to shake hands and form friendships with just about anyone. It also likes to link up with other carbons and make long chains. Long chains of carbon atoms that have hydrogens all along the sides are called **hydrocarbons**. Small hydrocarbon chains make things like natural gas (methane) and gasoline (petroleum). Medium-sized chains make things like wax, paraffin, and tar. Really long chains are found in plastics. Hydrocarbon chains can have other atoms attached to them, too, besides hydrogens. When chlorine joins the chain, we get PVC plastic (polyvinyl chloride) that is used for plumbing pipes.

Carbon also forms the "backbone" of many of the molecules in your body. Carbon atoms are the foundation, or anchoring points, for the other atoms in the molecules. Attached to the carbons, you'll find many of our non-metal friends: hydrogen, oxygen, nitrogen, phosphorus, and sulfur. In specialized bio molecules, you might find some metal atoms such as iron (in hemoglobin that carries oxygen) or zinc (in molecules that control DNA).   Carbon and its non-metal friends can be combined in almost endless ways, forming the vast number of biological molecules that make living things.

In this puzzle, write the letter symbol that goes with the atomic number written under each blank. For example, for the number 6 you would write the letter "C" for carbon. The letters will spell out the name of a substance that has carbon-based molecules. (Some letters don't appear by themselves on the table, so they have been written in.)

1) __ __ __ __ __
   15 57 16 22 6

2) __ __ __ __ __
   31 16 8 3 10

3) __ __ __ __
   59 8 91 10

4) __ __ __ __ T
   33 15 1 13

5) __ __ __ L
   13 27 67

6) __ __ __ __ __   __ __ L
   5 92 6 19 39   5 13

7) __ __ __ E __ __
   20 9 9   53 10

8) __ __ __   __ __ D
   95 49 8   89 53
   (building block of proteins)

9) __ __ __
   9 85 16
   (a type of wax)

10) __ __ __ __ __
    91 88 9 9 49

11) __ L __ __ __ R
    84   39 99 52

12) G __ __ __
     71 27 34
     (a sugar)

13) __ __
    105 77

14) G __ __ __
      71 52 7

15) __ __ L __ __
    7 39   8 7

16) __ __ __ __ __ __ __
    18 22 9 53 6 53 13

    __ __ __ __ RS
    9 57 23 8

17) __ __ __ T   __ __ __ R
    15 57 7   9 53 4

> Carbon can hold hands in 4 places!

> Carbon holding 4 H's

> Carbon is always drawn in black.

TRIVA QUIZ: What two letters of the alphabet do not appear in any of the symbols on the Periodic Table?

55

Activity 6.5 Practice makes perfect!

Here is a review activity to jog your memory about what you learned in past chapters.

1) If an atom could be enlarged to be the size of a sports stadium and the nucleus was sitting in the middle of the field, about how big would the nucleus be?
a) the size of a watermelon   b) the size of a marble   c) the size of a car   d) the size of an elephant

2) What do you call an atom that has more electrons than protons, or more protons than electrons?
a) an alkali   b) an isotope   c) radioactive   d) an ion   e) covalent

3) What is the valence number for oxygen? a) -2   b) -1   c) 0   d) +1   e) +2

4) What "family group" on the Periodic Table is perfectly happy? _____

5) What "family group" has only 1 electron in their outer shells? _____

6) What "family group" has 7 electrons in their outer shells? _____

7) Which element causes the stink in skunks and garlic? _____

8) Which element can form a circle called a buckyball? _____

9) Which element is taken from the air by bacteria and put into the soil? _____

10) Which element was first discovered in the sun? _____

Match the formulas with their common names. (Word bank: plaster, sand, Teflon, salt, bleach)
11) SiO_2 _____
12) NaCl _____
13) NaClO _____
14) C_2F_4 _____
15) $CaSO_4$ _____

"We know all the answers but we're not not telling!"

16) An atom of magnesium is most likely to bond with: a) N   b) C   c) O   d) F   e) Ne

17) An atom of potassium is most likely to bond with: a) Na   b) B   c) Ca   d) Mg   e) Cl

18) The atomic number is the number of _____ that an atom has.

19) Which of these things is NOT made of carbon?   a) diamonds   b) graphite   c) coal   d) glass

20) Which of these statements is NOT true about electrons?
a) Electrons don't like to be close to each other.   b) Electrons like to "work" in pairs.
c) Electrons have a positive electrical charge.   d) Electrons weigh almost nothing.

BONUS QUESTIONS (a little harder)

1) What atom is this? $1s^2 2s^2 2p^6 3s^2 3p^4$? _____

2) Protons and neutrons have a mass (weight) of 1 amu (atomic mass unit). The mass of an atom is equal to the number of protons plus the number of neutrons. If an atom of uranium has a mass of 238 and uranium's atomic number is 92, then how many neutrons does this atom have? _____

3) When you see a number outside of the parentheses, like the 2 in this formula: $(OH)_2$, that means you have two of whatever is inside of those parentheses, in this case 2 (OH)'s. So you have 2 O's and 2 H's.
How many O's (oxygens) are in this mineral? _____ $Ca_{10}Mg_2Al_4(SiO_4)_5(Si_2O_7)_2(OH)_4$

CHAPTER 7: METALS: SEMI-, PURE, AND TRANSITION

Working our way to the south and the west in our Periodic Kingdom, we come to the semi-metals. They live along the diagonal line, in between the metals and the non-metals. In our fairy tale, we made up that part about the non-metal family having a difficult last name that no one could remember. It just sounded like a logical reason for them to be called non-metals, and it made the story more interesting. But it is true that the semi-metals can also be called the metalloids. You will see both names used equally.

The metalloid neighborhood is anything but a settled place. Chemists don't all agree about exactly where the dividing line is. Some say boron should be a metalloid, others say not. Some say polonium should be a metalloid, others say it is a true metal. So don't be overly concerned about trying to remember which are which because chemists don't even know for sure! In this booklet, we'll just choose the pattern that is easiest to remember. We are going to use a stair-step right down the diagonal, then add two squares under the middle. Looks nice, eh?

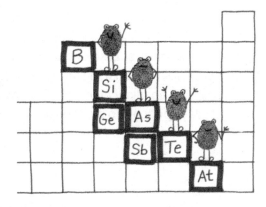

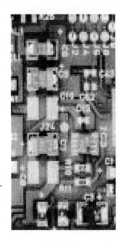

The only metalloid we are going mention in particular is silicon. (The others are used in high-tech industries in various ways, and astatine is radioactive—more about that in the next chapter.) Have you ever heard of "Silicon Valley" in California? You might think that's where you can find a lot of silicon in the soil, but the name actually comes from the industry that goes on there: computers and microchips. Because silicon is on the borderline between metals and non-metals, it can act like both. Sometimes it can act like a metal and be very good at carrying electricity, but in other situations it acts like a non-metal and be an insulator that doesn't carry electricity. Silicon is the ideal element for making some of the electronic parts that are used in computers and other high-tech equipment. Silicon Valley is an area that has a large number of high-tech computer companies. The element silicon has come to represent all computer-based technology.

Silicon is also one of the most abundant elements on the planet. It is one of the main ingredients in many kinds of rocks. Volcanic lava is high in silicon, so when it cools, the rocks it forms are high in silicon. We've already seen that the chemical recipe for sand is SiO_2. Strangely enough, when small amounts of other elements are mixed in with SiO_2, it isn't ruined. The impurities turn it into a variety of semi-precious gemstones, such as jasper, agate, opal, amethyst, onyx, and chalcedony. The purple mineral shown here is amethyst and below it is an agate.

a diatom

In the world of biology, silicon is used by some animals, such as diatoms, to make their hard outer shells. Others, like sea sponges, use silicon for their structural skeletons. (Most shelled animals use calcium instead of silicon.)

purple amethyst and striped agate

The true metals have only three members that most people have heard of: aluminum, tin and lead. The others, gallium, indium, thallium, bismuth, and polonium, are not well-known. (You've probably heard of bismuth without being aware of it. The "bis" in "PeptoBismol" stands for bismuth. Bismuth is one of the key ingredients!) Before you started reading this book, if we had asked you to name some true metals, you would probably have included elements such as copper, nickel, iron, silver and gold. It's surprising that these elements we know as metals are not in the true metal family on the Periodic Table. When chemists named the groups, they were looking primarily at the electron arrangements, not at how the elements are used. The "true" metals are "true" because of their electron configurations and, therefore, their location on the Table. Here we have another case of chemists using common words in a different way. We saw this with the word "salt." To a chemist, a salt is not something you put on food. It's what you get when you combine a halogen with an alkali element. Similarly, chemists use the word metal in a way that is different from our everyday speech. To a chemist, most of the elements on the Periodic Table are metals.

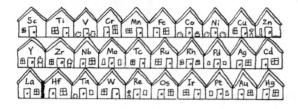

Next we come to the very large neighborhood right in the middle of the table, where all the transition metals live. In our story, they were the hard-working people of the kingdom who worked at industrial jobs. Many of the elements in this block have familiar names, such as titanium, chromium, iron, cobalt, nickel, copper, zinc, silver, cadmium, tungsten, platinum, gold, and mercury. Others are strangers to you, such as yttrium, niobium, molybdenum, osmium and iridium. You might look at some of their names and be afraid to pronounce them. One way to become comfortable with things that seem hard is to introduce them in ways that seem amusing. So here's a funny introduction to five of these strange transition metals.

The strangest neighbor on the block is technetium *(tek-NEE-shee-um)*. His next-door neighbors report that they've seen harmful radioactivity coming out of his house. And most of the time he isn't even home. Everyone says he doesn't belong in this neighborhood because no one else is radioactive, but when he shows them his electron configuration, he does indeed fit right between molybdenum *(moll-LIB-den-um)* and

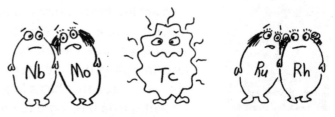

Molly leans over to gossip with Mr. Niobium, and Ruth whispers to her friend, Rhoda.

ruthenium. Poor Molly and Ruth! Ruth spends a lot of time gossiping to her neighbor, Rhoda, about it. Before technetium moved in, he lived at the nuclear power plant. He said that's where he was born. Fortunately for the neighbors, he's not around most of the time.

Technitium's stable neighbors, niobium, molybdenum, ruthenium and rhodium are often combined with other metals in order to improve their chemistry for making things like tools, heaters, bulbs, lasers, and spark plugs. Unlike its neighbors, technetium is a synthetic (man-made) element and is often a by-product of nuclear fission in a nuclear power plant. Since it is man-made, the number of neutrons in its nucleus can be controlled to some degree. Technetium atoms with 56 neutrons are less dangerous and their radioactivity wears off in about 6 hours. This makes them suitable for use in medical applications such as doing scans to determine areas of disease. (Molly won't want to admit this, but sometimes atoms of molybdenum are used to make this type of technetium. How can this be? Remember, it's the number of protons that defines an element. Add a proton to Mo, and you get Tc.)

There are so many transition metals that we can't discuss each one. Fortunately, you are probably already at least somewhat familiar with many of them, such as gold, silver, iron, copper, nickel, zinc, platinum and mercury. Others, such as tungsten, might have unfamiliar names, but we interact with them all the time without even knowing it. Tungsten is what those thin filaments inside light bulbs are made of. Cadmium is used in rechargeable batteries, and we see chromium on the surfaces of shiny tools and car parts. Vanadium is an ingredient in the metals that are used to make wrenches and pliers.

When two or more metals are mixed together we call this mixture an **alloy**. One of the first alloys ever discovered was **bronze**. Ancient metal workers found that when they added some tin to their molten copper, the result was a metal that was much better than just plain copper. Bronze was harder and more durable than copper, making it better for weapons and statues. Later, during the Roman period, zinc was added to copper to make **brass**. Brass was even more durable than bronze, and if you added some aluminum or iron to the mix, the resulting metal was very resistant to corrosion and could be used to make parts for boats that were constantly exposed to salty ocean water. Metal workers over the centuries tried adding tiny amounts of various other elements, such as arsenic, phosphorus or manganese. Each element would have an effect on the final product, letting them adjust the quality of the metal to suit the application it was being used for. One of the brass alloys is ideal for making musical instruments such as trumpets and trombones. Another variation of brass is used to make cymbals.

a bronze statue

A volatic pile similar to the ones that Humphry Davy used to discover sodium. (Image: Wikipedia, GuidoB.)

Metals have another important characteristic, besides being durable— they can carry an electrical current. Some metals are better than others at conducting electricity, but all metals are at least somewhat conductive. The very best conductors are silver, copper and gold. Since copper is much less expensive than silver and gold, it is the best choice for electrical wires. Nickel, zinc, iron, cobalt and tin are in the middle, and lead is at the bottom of the list. (The transition metals we have not mentioned are somewhere in the middle, too, but listing all of them is a bit much. If you want to see a list, you can always consult the Internet.)

An Italian scientist named Alessandro Volta figured out how to use copper and zinc to produce electricity. He made a stack of three types of discs: copper, zinc and leather saturated with salt water. The salt water discs between the copper and zinc allowed the electrical current to flow to the next set of discs. Voltaic piles were fairly easy to make, and soon many scientists were making them. One of those scientists was Humphry Davy, who used as many as 100 piles hooked together to make a dangerously strong current. Davy's electricity was strong enough to isolate pure sodium atoms. But how did these piles work? What was happening inside the metals? It wouldn't be until the 20th century that the answer was discovered.

A copper wire is made of millions of copper atoms stuck together, so obviously copper atoms stick to other copper atoms. They don't use ionic or covalent bonding, though. The transition metals use their own type of bonding called **metallic bonding**. When these elements get together, they don't keep track of their own personal electrons very well. Their electrons are free to wander about. It's sort of like a parent saying to a child, "You can play anywhere in the neighborhood, just don't leave

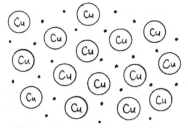
Copper's outer shell electrons are free to move about.

61

the neighborhood." The child might go next door to play in his neighbor's yard, or he might wander down the street a bit. However, in the atomic world (unlike the real world) you can't rule out the possibility that the electron "child" might actually wander off permanently. Electrons are not individuals like humans are. All electrons are the same. If a copper atom's outer electron wandered off and another one came to take its place, the copper atom would never know the difference. Will the copper's electron stay in the area? Probably. Will it leave? Maybe.

While it is true that electricity is made of moving electrons, we shouldn't think of a stream of electricity as being like a stream of water. Rather, a better analogy might be a row of dominoes. In a domino rally, the dominoes move and we see a pattern of motion being carried along from one end to the other, but the dominoes basically stay in the same place. Or, you could think of a row of people standing in a line. The first one in the line pushes the second one, who pushes the third one, and so on down the line. (The people would not have to be reset like the domino rally would, so perhaps the people are a slightly better analogy.)

Look at the Periodic Table and you will see that copper, silver and gold are all in the same column, with copper on top, silver in the middle and gold on the bottom. Remember, the columns tell us about the arrangement of the electrons in the outer shell. For example, all the elements in the noble gas column have full outer shells and all the halogens have one less than they'd like. The Quick-And-Easy Atomizer activity showed us that the placement of electrons up to argon is fairly straightforward. If we had kept going, however, it would have gotten messy. With the transition metals, things are

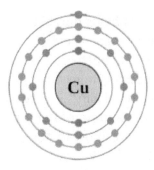

An over-simplified drawing of copper's electrons. The s, p and d orbitals are not shown.

not so straightforward. The addition of the 10-seater "d" shell makes things more complicated. Every time an electron is added, the "math" changes. Transition metals can decide how to split up the electrons into the s, p and d orbitals in a way that maximizes electron "happiness" (or at least minimizes "unhappiness"). The diagram shown here is the one you will see if you Google "copper electrons." It shows you the end result of what copper has done with its electrons, but it does not show you the s, p and d orbitals. That one outer electron is actually sitting in the 2-seater s orbital, which should have been filled first, according to our Atomizer rules. Copper decided that it was better to have a half-filled 2-seater orbital than to have a 10-seater orbital with one electron missing. Copper likes the fraction "1/2" better than "9/10."

Having just one electron in the outer shell doesn't make these elements act like alkalies, though. Copper, silver and gold certainly don't explode when you put them in water! Quite the opposite—they are very stable. In the transition neighborhood, having one outer electron makes you very good at conducting electricity.

One last characteristic of transition metals really needs to be mentioned. If you start peeking into higher level chemistry books, it won't take you long to discover that many of these elements have multiple valencies (which are usually called "oxidation states"). Chromium, for example, has three options: (+2), (+3) and (+6). Chromium's oxidation number (valency) depends on what atom, or atoms, it is bonding with. It's the same with copper and iron and many other metals. Their oxidation states can change so that they can bond with atoms such as oxygen, nitrogen, chlorine, or carbon. This makes learning chemistry a lot more difficult, but it also allows for a larger variety of minerals to exist. Our world is more beautiful, more diverse and more interesting because transition metals have more than one oxidation state (valency).

A colorful opal
from Wikipedia (Dpulitzer)

Activity 7.1   "The Bonding Song"

Now that you know about all three types of bonding, you are ready for "The Bonding Song." There are two audio tracks, one with the words and one without, so that after learning how the words go, you can sing it yourself. (If you don't already have the audio tracks, you can access them by going to: www.ellenjmchenry.com/audio-tracks-for-the-elements)   The tune might sound familiar, as it was borrowed from the American folk song, "Turkey in the Straw."

The Bonding Song

There were two little atoms and they both were very sad,
They wanted eight e's but six was all they had,
Then they hit upon a plan and decided they would share:
They each gave the other an electron pair.
Covalent bond – sharing is great!
Covalent bond – now they both have eight!
Outer shells want eight electrons,
So non-metal atoms form covalent bonds.

There were two little atoms, they were sad as you might guess,
They wanted eight e's, one had more and one had less,
Then they hit upon a plan and one atom said,
"I'll give my extra e's if you agree to wed."
Ionic bond – one atom gives!
Ionic bond – one atom gets!
Atoms give and take their electrons
And they stay right close together with an ionic bond.

There were lots of little atoms, they were metals every one,
They were all in a clump, they were having lots of fun,
And the way they stuck together was to share their e's around;
They called their little clump a metallic bond.
Metallic bond – everyone gives!
Metallic bond – everyone gets!
Electrons float and belong to everyone,
And the metals stick together with metallic bonds.

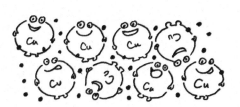

Activity 7.2   An online quiz about metals

Here's a just-for-fun quiz game you can do to test your knowledge of metals. If you don't know the answers, just guess—you'll still learn! (This link has been stable over the years, but if you find that it doesn't work, try searching the Internet with key words "online quiz metals.")

http://www.syvum.com/cgi/online/serve.cgi/squizzes/chem/metals.tdf?0

Activity 7.3   Finish memorizing the Table up to radon

Why in the world would you want to do this? Well... if you are going into chemistry some day, the answer is obvious. If not, it's still kind of a cool thing to do and a great way to impress your relatives at holiday gatherings. If nothing else, it's good exercise for your brain, like lifting weights is good for your muscles. However, it is also optional. If you've had enough memorizing, that's okay.

If you want to try it but feel like you need some help, you might want to use a "mnemonic" (the first "m" is silent). A mnemonic is a story, a picture, or even a silly idea, that helps you to remember something. It's not cheating, it's just being clever about helping your brain to make information stick better. Here is a mnemonic story about the next two lines on the table, in case you find it helpful, (and yes, it is very silly), but you could also make up your own.

RUBy was the STRongest woman in her town. Y, her muscle "Z-R" the "N-B"
rubidium *strontium*   "Why" *yttrium* "they are" *zirconium* "envy" *niobium*

of everyone. Her friend, MOLLY, claims Ruby won the river race fair and square.
molybdenum

TECHnically, RUTH, RHOD faster, but she was a PAL and AG-cepted the silver
technetium *ruthenium* *rhodium* "rowed" *palladium*

medal instead. Before the race, a CAD named INDY, took a pair of tin SNips,
cadmium *indium*

ANT SaBotaged her boat. She yelled at him, "I can TELLUR a cheater!"
"and" *antimony = Sb*   "tell you're" *tellurium*

She punched him and they had to put IODINE on the cut.   Queen XENON

sent Indy to the dungeon.

Once there were two brothers, CESIUM and BARIIUM. Their science homework

was to research LANTHANUM and the LANTHANIDE series. Cesium said,

"We'll HAF-TA check out the WWW. You can REally get a lot of information there."
"have to" *hafnium tantalum* "World Wide Web" *tungsten=W* "rowed" *rhenium*

Just then, their dog, OSMIUM, came in. "IRrr," he growled. He had an empty
iridium

PLATe in his mouth. "A-U HoG!" shouted Cesium. "You can TL he ate our PB
platinum "Hey, you hog!" *Au=gold Hg=mercury* "tell" *Tl=thallium* "peanut butter" *Pb=lead*

sandwiches!" Barium said,"That's okay, we can always BI more. Or we can eat
"buy" *Bi=bismuth*

POLOgna istead." AT last, we've made it to RADON!
"bologna" *Polonium*   *At = astatine*

64

Activity 7.4 Who am I? (A guessing game about <u>pure metals</u> and <u>semi-metals</u>)

1) _____ I have a very low melting point. This means I might even melt in your hand. I was named after the old Latin word for France.

2) _____ Although I am famous for being poisonous, many living things (including humans) need me in very small amounts. When combined with gallium I am an important ingredient in electronic devices.

3) _____ My letter symbol does not match my name. I used to be used for water pipes, but not any more. I am very dense, which makes me feel heavy.

4) _____ I used to be called stibium. In the ancient world, I was used in cosmetics. Now I am mainly used in fire-proofing and in lead-acid batteries.

5) _____ My name means "indigo blue" because I have a bright blue line in my spectrum when I am burned. I am similar to my Periodic neighbors and have a low melting point, making me useful for soldering. I am also used in high-tech products such as semiconductors.

6) _____ I am best known for my alloys. If you add me to copper, you get bronze. I used to be made into cans, but now they use aluminum instead.

7) _____ I sit right next to a liquid element, but I am not a liquid. My name means "bright green twig" because I have a bright green line in my spectrum. I am used in high-tech industry as an ingredient in sensors and detectors, but don't eat me because I am poisonous.

8) _____ I sit next to many toxic metals, but I am not poisonous. In fact, I am used in medicines and cosmetics. No one knows where my name came from, but perhaps from the Arabic word for antimony. I was often confused with antimony since we have many similarities due to the fact that we are in the same column on the Periodic Table.

9) _____ My name comes from the Latin word for "earth." I am very rare so I am used in small quantities. I am added to copper and lead to make alloys that are more easily "machined" than they would be otherwise. I have some chemical similarities to selenium and sulfur.

10) _____ I am the third most abundant element in the earth's crust. I am one of the least dense elements, making me feel very light. I am not magnetic at all, but I do conduct electricity.

Activity 7.5   Research your favorite transition metal

Choose a transition metals that you'd like to know more about. Use the following page to record the findings of your research. If you'd like to research more than one transition element, you can make an extra copy (or copies) of the page before you write on it.

_____

name of element

atomic mass	number of protons	number of neutrons	number of electrons

symbol

atomic number

At standard temperature and pressure (STP), this element is a:

☐ solid
☐ liquid
☐ gas

Where did this element get its name?

At what temperature will this element boil?

At what temperature will this element melt or freeze?

What group does this element belong to?

☐ alkali metals
☐ alkali earth metals
☐ transition metals
☐ true metals
☐ semi-metals (metalloids)
☐ non-metals

Is this element found in the Earth's crust? ☐ yes ☐ no
If so, where?

☐ rocks ☐ dirt ☐ lava
☐ water ☐ gemstones
☐ sand ☐ _____

When was this element first discovered?

Who discovered it?

Is this element ever found all by itself (not part of a compound)? ☐ yes ☐ no

What color is this element? (Or, if it is never found by itself, what color is its most common compound?)

Other colors?

Is this element used in industry? ☐ yes ☐ no
If so, what is it used for?

Is this element found in the human body? ☐ yes ☐ no
Is it part of the structure of the body? ☐ yes ☐ no
Can this element be harmful to the body? ☐ yes ☐ no
If it is harmful, how might you ingest it or come into contact with it?

Is this element used in any art or craft? ☐ yes ☐ no
What type of art uses it?
☐ painting ☐ sculpture ☐ pottery
☐ printing ☐ coins ☐ jewelry
☐ _____

Is this element used in medicine or dentistry? ☐ yes ☐ no
If so, how is it used?

Give one historical fact about this element other than the date of its discovery:

Draw a picture of a molecule containing this element:

What do you think is the most interesting fact about this element?

Name of molecule:

68

CHAPTER 8: THE LANTHANIDES AND ACTINIDES

The lanthanide and actinide series are almost always shown as two separate rows, sitting below the main table, as if they were not really part of the table. Actually, they <u>are</u> part of the table, and if we put them in where they belong, the table would look like this:

The problem with this table is just practical—it doesn't fit very well on a single page. By the time you shrink it down enough to get it onto the page, the squares are so small that you can't read the information written in the boxes. It's so long that it's awkward. Taking out the longest, skinniest part, the lanthanide and actinide rows, makes the table look much better. The crack in the street in the Periodic Kingdom is where these two rows should go.

What are these two rows? In the Periodic Kingdom the lanthanides were industrious miners who provided rare metals for high-tech products. The real science is very close to this picture!

These elements were once thought to be very rare, although we now know they are more abundant in the Earth's crust than silver or gold. We can still call them rare, though, because you don't find them sitting around in their pure form waiting to be collected. No one has ever gone panning for neodymium like they would for gold. You don't find chunks of solid cerium when digging with a shovel, as you might with copper. These rare earth metals are more difficult to get out of the rocks in which they are found because they are mixed in with so many other things.

A rock that contains minerals is called an "ore." This picture shows an ore that contains rare earths. The penny sitting on it lets you know how big the rock is. Miners dig up the ore, then have to figure out how to get the elements out of the ore. This usually involves crushing and heating the rocks, plus the addition of chemicals. Only after a lot of work is the pure element obtained. The rare earths present a problem for the refiners because the temperature at which they melt is very high. However, it can be done, and, in fact, every year millions of tons of rare earth elements are mined and refined. (China produces 95 percent of the world's rare earths.)

69

The rare earth elements would have been useless to ancient peoples. They're not good for making pottery or jewelry or weapons. In the modern world, the rare earth elements are used in many "high-tech" products, such as color televisions, computer screens, lasers, cell phones, solar panels, spark plugs, camera lenses, x-ray screens, mercury lamps, lasers, medical imaging film, temperature-sensing optics, nuclear reactors, self-cleaning ovens and welding goggles. Without rare earths, "green" technology would not be possible. You can't have solar panels and wind turbines without mines that dig up and process rare earth ores.

These powdered rare earths were obtained from ores. They are not pure elements. Pure rare earths look shiny.

What makes rare earth elements so useful for technology? Basically two reasons: their **magnetism** and their ability to **fluoresce** *(flor-ESS)*.

Magnetism occurs when most of the electrons in a substance are spinning the same way. Remember those rules that electrons live by? The first rule was: "Spin!" When electrons pair up

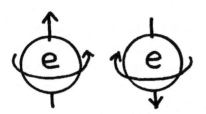

(another rule they live by) they always choose a partner who is spinning the opposite way. Pairing up with their opposite "neutralizes" their spin, so the more <u>un</u>paired electrons an element has, the more likely it is that the element will be magnetic. The elements in the middle of both the transition neighborhood (such as iron and cobalt) and in the lanthanide row (from neodymium to gadolinium) have the greatest number of unpaired electrons, so they are the most magnetic elements. (Promethium might be magnetic, but because it is radioactive it is not suitable for use in magnets.) How many unpaired electrons can a rare earth element have? The outer shell is an "f" shell, which can hold up to 14 electrons arranged into 7 pairs. The first 7 electrons get their own "seat" in the shell, then after that, the additional electrons have to start pairing up. So the maximum number of unpaired electrons is 7. This corresponds to the element gadolinium.

Neodymium and samarium might not have the maximum number of unpaired electrons, but they actually turn out to be the most useful for making magnets. On their own, however, their magnetism occurs only at low temperatures. They are mixed with transition metals such as iron, nickel or cobalt, which are magnetic at higher temperatures. One of the most common alloys of neodymium

A bracket containing a neodymium alloy magnet from the hard drive of a computer.

also has some boron in it: $Nd_2Fe_{14}B$. The crystal structure of this compound also happens to be very good for magnetism, so between the unpaired electrons and the crystal structure, this compound is so magnetic that even very tiny magnets are extremely powerful. This is handy for the electronics industry because if they had to use regular iron-cobalt magnets, cell phones would be ten times larger, and who would want to carry those around?! Computer hard drives also contain neodymium alloy magnets—another item where small size is a definite "plus."

Wow! These lanthanides turn out to be really important!

Not all rare earth alloy magnets are small. You will also find them in cordless tools, electric cars, loudspeakers, headphones, and MRI machines in hospitals. Perhaps the largest objects that contain rare earth magnets are the generators inside the huge wind turbines used to create electricity from wind. Looks like if you want green energy

Most of the rare earth elements also **fluoresce**. You see fluorescence *(flor-ESS-ence)* every time you look at a fluorescent bulb. (Fluorescent bulbs are usually very long or are spiral-shaped.) Magic markers used for highlighting textbooks also fluoresce. Laundry soaps often have fluorescent dyes that fool your eyes into thinking that the whites are "whiter than white." Other things fluoresce to some degree, but not enough for your eyes to be able to see it very well.

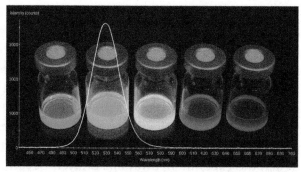

These liquids fluoresce when exposed to UV light. They are shown on a graph that gives the exact frequencies of light at the bottom. The spike in the graph corresponds to the intense glow of the green liquid. The others glow less, which is shown in the graph as a very low line.

Fluorescence is caused by "falling" electrons. When an electron (usually in the outer shell) gets "zapped" with some extra energy (ultra violet light, for example) the electron is "excited" into a higher energy state and "jumps" up to the next higher shell. However, it can't stay in that higher shell forever. Just as you must come back down when you jump into the air, so an electron can't stay at a higher energy level and must come back down to its normal level. When it falls back down, it releases the energy that it had absorbed. Often, the energy that is released doesn't look or act the same as the energy that went in. For example, when UV (ultra violet) light hits some atoms, the released energy isn't UV, but is a visible light such as green or red or blue. Scientists like to hit atoms with X-rays and watch what happens. X-rays are particularly useful in helping to figure out the molecular shape of a crystal or the identity of a mystery atom in the crystal. Heat can also cause electrons to jump. Think back to the discovery of helium. The scientists saw a striped pattern of colored lines in their spectrometer because the sun's heat was causing electrons in helium atoms to jump up and down, releasing their energy as bands of visible light. Each element has a unique pattern of electrons, and therefore a unique pattern of emitted light.

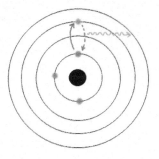

"What goes up must come back down." That includes you!

You may be wondering how an electron in an outer shell can jump to a higher shell. Is there an empty shell sitting around the outer one? Basically, yes. ALL the shells exist in every atom, whether they are filled or not. Imagine that each atom has an Atomizer pattern printed around it (but with many more orbitals than in our Atomizer activity). The rings are waiting for electrons to come and fill them. The outer rings of small atoms never get filled, of course. Large atoms such as uranium and plutonium use all, or at least most, of their rings.

There are empty levels above the outer electron shell. One electron has jumped and then fallen back down, emitting light.

Now, back to the rare earth elements. **Europium** can fluoresce with either a red or blue light, depending on what other atoms are surrounding it. **Terbium** fluoresces bright green. These two elements are the key ingredients in making the colors you see on televisions and computer screens. Believe it or not, red, blue and green light can be combined to make any color—even yellow, brown, black and white.

Activity 8.1   Look at a screen

You will need a magnifier for this activity (at least 10x). Look at a computer or television screen while it is on. If your magnification is high enough, you will see that the image is nothing but red, green and blue dots or rectangles. Choose a place where you think it looks white or yellow or brown, then zoom in again. Is there really any white or yellow or brown? How can red, green and blue make white? Amazing!

Our Periodic Kingdom story ended with "Beware the Actinides!" because the elements in this row are all **radioactive**. Radioactivity was first discovered by a French scientist named Henri Becquerel in 1896. He was experimenting with rocks that were fluorescent and phosphorescent. (**Phosphorescence** *(foss-for-ESS-sense)* is when the electrons keeping falling for a number of minutes, causing the "glow in the dark" phenomenon.) Becquerel would cover a photographic plate with black paper then allow sunlight to strike the rocks, which would fluoresce. He hoped the fluorescence would go through the black paper and make an image of the photographic plate. When weather turned cloudy one day, he put the experiment away in a drawer. Several days later he opened the drawer and was shocked by what he saw. The photographic film had a very clear image of the rocks on it. How could that have happened in a dark drawer without any light? Becquerel correctly guessed that something in the rocks was giving off a type of energy that did not depend on sunlight. He knew that the rocks contained the element uranium, and supposed that it might be the uranium that was doing this. Becquerel then handed off the investigation to Maire Curie.

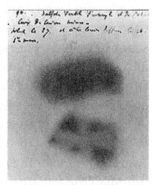

This is Becquerel's image of the rocks in the drawer.

Marie Curie in her lab in Paris.

Marie Sklodowska had come to France from Poland. While studying and working in Paris, Marie and met and married Pierre Curie, another brilliant scientist who was discovering many new things. Marie began studying a mineral ore called **pitchblende** that was known to contain uranium. She boiled samples of pitchblende for several years, trying to get all the uranium out. After the uranium was gone, the remaining minerals still gave off rays of energy. This led her to believe there was another element still in the rock that was capable of producing the same rays of mysterious energy. Eventually she managed to isolate a small amount of a new element. She decided to name it **polonium** after her native country, Poland. She named this new type of energy **radioactivity**.

After the polonium had been extracted from the rocks, they STILL gave off radioactivity! Could there be a third radioactive element? Sure enough, there was. When she finally isolated this third radioactive element, she named it **radium**, and calculated that it was 200 times more radioactive than uranium. Good thing she'd been wearing protective gear while... oops—she hadn't been wearing any protection at all. No one at that time had any idea how harmful radiation was. They thought it was just an interesting thing that some elements did. Amazingly, Marie lived until the age of 60 before she died of a cancerous illness caused by the massive amount of radiation she had been exposing herself to for years.

Most of the chemical properties of an atom come from the electrons in its outer shell. Radioactivity, though, is all about the nucleus.

The atomic mass of [the most common form] of uranium is 238. That means if you add up all the protons and all the neutrons, you will get a total of 238 particles. That's a huge nucleus! In fact, it is so huge that it has trouble staying together. It's like a big crumbly cookie; little bits can break off easily. When an atom's nucleus

begins to crumble, it doesn't drop crumbs or chocolate chips, of course. It flings out protons, neutrons, and rays of dangerous energy. "Radioactivity" is the name for those particles flying out of the nucleus.

The particles that are ejected from the nucleus are classified into three groups: alpha, beta, and gamma. Those are the first three letters of the Greek alphabet, so it's like saying A, B and C. Alpha particles consist of two protons and two neutrons. Beta particles are basically high energy electrons, and gamma particles are made of the same kind of energy as light (electromagnetic energy) but carry a dangerously high level of that energy. Let's focus on alpha particles for a few minutes because they turn out to be part of the process of one element turning into another. (The alchemist's dream comes true!)

An alpha particle is made of two protons and two neutrons. If this particle can capture two electrons, it will then have exactly the same structure as a helium atom. Electrons are pretty much everywhere all the time, so it is not hard for an apha particle to find two electrons. Once it has captured the electrons, it is no longer an alpha particle but has become a genuine helium atom. This is what eventually happens to all alpha particles—they become helium atoms. So when geologists find helium atoms inside rocks or between rock layers, they assume that there was once radioactivity in that area.

These symbols are used as warnings about radioactivity.

What happens to a uranium atom if it ejects an alpha particle? The nucleus no longer has 92 protons; it now has only 90. Since the number of protons defines what element an atom is, the uranium is no longer uranium, but has turned into element

An alpha particle that has captured two electrons and turned into a helium atom

number 90, thorium. Thorium is still radioactive, however, and will eventually eject an alpha particle. When it does, the nucleus will no longer have 90 protons, but will have only 88. Thorium has turned into element 88, radium. If this happens again, radium will lose two protons and turn into radon, 86. Radon is radioactive and could lose two protons and become element 84, polonium. Polonium is also radioactive, as Marie Curie found out, and is capable of giving off an alpha particle. If polonium loses two protons, it will turn into 82, lead. Lead has a stable nucleus and is not radioactive. Lead won't turn into anthing else. Lead is a stopping point for this process that we call radioactive "decay."

The entire actinide row is radioactive, but as we've seen already, there are other elements on the table that are radioactive, as well. Starting with 84, polonium, all the higher elements are radioactive. There are also two radioactive elements with numbers less than 84: technetium, 43, and promethium, 61. So we can say that all actinides are radioactive, but not all radioactive elements are actinides.

Uranium is generally considered to be the last **naturally occurring** element on the Periodic Table. (We must say "generally" because recently there have been reports of very tiny amounts of neptunium being discovered in uranium ore. However, most science books still list uranium as the last naturally occurring element.) Plutonium and the other actinides can't be dug out of the ground. They simply do not exist anywhere in nature. They are completely man-made elements. Man-made? Can elements be manufactured? Yes, if you have a nuclear reactor and a particle accelerator.

We just saw the process of radioactive "decay," where elements lost pairs of protons and changed into lower elements, all the way down to lead. What would happen if you could <u>add</u> protons? Because elements are defined by the number of protons they have, if you added a proton to uranium, it would no longer be uranium. Until the 1940s, there was no way to add protons to atoms. Then, during World War II, scientists in both Germany and the USA discovered how to make machines that would shoot protons at atomic nuclei. Unfortunately, the reason they wanted to do this was to make nuclear bombs. However, aside from bombs,

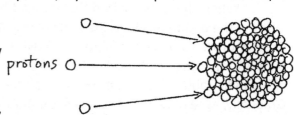

protons

this technology proved to be useful because it could allow chemists to make many new elements. They would shoot protons at an atom that was already very large, such as thorium or protactinium or uranium, and hope that some of the protons would stick to the nucleus. They were very careful

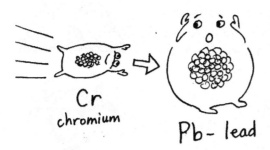

Cr
chromium

Pb - lead

Sometimes two medium-sized atoms are crashed together in the hopes that they will stick and make a super heavy element. To make 106, chromium (24) and lead (82) were used. The math works: 82 + 24 = 106

about their technique so that they knew exactly how many protons were sticking. As the number of protons grew, so did the number of new elements. Each time they got one more proton to stick, they created a new element. These new elements needed names, and this ended up being a way to honor the memories of great scientists such as Marie Curie, Albert Einstein, Enrico Fermi, Dmitri Mendeleyev, and Alfred Nobel. Some of the new actinides were named after the places where they were created, such as Berkeley, California, USA (berkelium, californium and americium), hassium (Hesse, Germany), dubnium (Dubna, Russia) and hafnium (Hafnia, the old name for Copenhagen, Denmark).

A big problem with these super heavy elements is that their nuclei are too large to ever be stable. They are doomed to fall apart eventually. However, some super heavy nuclei are stable enough to stay in existence for weeks or months, long enough that scientists can observe them and determine some of their chemical properties. They always turn out to be similar to the elements that are above them in the same column. Other super heavy elements fall apart after a few seconds, or even a fraction of a second. They wink out of existence before any tests can be done on them, so we know much less about them.

Scientists made so many new elements that they filled up the actinide row and had to go back to the bottom row of the transition block (the ghost

town in the Periodic Kingdom, where the residents are rarely seen). When someone claimed to have "discovered" a new element, it would take a long time, perhaps even years, to verify that they had actually done so. Experiments and data would have to be published so that scientists all over the world would be able to read about what had been done and agree that a new element had indeed been made. Only after the discoverer's claims were judged to be valid would a name be chosen for the new element.

Naming an element is often complicated by disagreements about who was first to make it. For example, Russia and America argued over element 106 for a long time before the international chemistry naming group, IUPAC, decided that America had been first. Whoever discovers an element gets to name it, so, of course, the Americans choose to name it after an American scientist. They chose Glenn Seaborg, who has been given credit for discovering, or helping to discover, plutonium, americium, curium, berkelium, californium, einsteinium, fermium, mendelevium and nobelium. (As you might guess, Seaborg's lab is located in Berkeley, California.)

When chemists realized that it might be possible to make a lot more new elements, they began labeling the empty blocks in the bottom row with temporary, fictional names made from Latin and Greek words. For example, before 106 was named, it was "unnilhexium." ("Un" is Latin for "one," "nil" is Greek for "zero," and "hex" is Greek for "six.") You will still see at least four temporary names on all Periodic Tables. Their abbreviations all start with the letter U, such as UUT, UUP, UUS, and UUO. Gradually, as real names are chosen, the temporary names will disappear.

Activity 8.2   A video about Marie Curie and the discovery of polonium and radium

Marie Curie discovered polonium in 1898 and then radium soon after. Marie's life story is amazing and inspiring. There are several videos about her posted on YouTube playlist.

Activity 8.3   A virtual field trip

The professors who made the Periodic Table of Videos filmed a trip they made to Ytterby, Sweden, to find the famous mine that was the source of the mineral ore that allowed the discovery of Yttrium, Terbium, Erbium and Ytterbium.

The mnemonic story below will make more sense if you've seen this video first. Watch "Ytterby Road Trip" on The Elements playlist on the YouTube channel.

Activity 8.4   Memorize the Lanthanides and Actinides

This activity is optional, of course. But for those of you intent on memorizing the table, perhaps this little mnemonic story might help. (You can always make up your own story, too!)

The Lanthanides are for those who are SERIOUS about learning the whole table.
(cerium)

Begin by PRAISING NEODYMIUM for its wonderful magnetic properties. Then tell this
(praseodymium)   (Neodymium magnets are extremely strong.)

story: "I PROMised to take SAM to EUROPE so he could see the famous mine where
(promethium)   (samarium) (europium)

(This)
Johan GADOLIN discovered TERBIUM. When we got there, he said, "DYS HOLE isn't
(gadolinium)   (dysprosium) (holmium)

what I was expecting! I was expecting an URBan area with "inTHULated" buildings.
(the mine was a big disapointment)   (erbium)   (thulium)   (It's cold in Sweden!)

Let's leave YTTERBY and go back to the row with the LOOT! (meaning the row that has gold in it)
(ytterbium)   (lutetium)

For the Actinide Series, you need to be THORoughly PROTECTed because they're all
(thorium)   (protactinium)

radioactive. URANIUM, NEPTUNIUM and PLUTONIUM are the most famous members

of this row because they were used by AMERICA to make atomic bombs. Marie CURIE
(americium)   (curium)

would have loved to visit BERKELEY, CALIFORNIA to see the scientists make
(berkelium)   (californium)

EINSTEINIUM and FERMIUM. But she, MENDELEYEV and Alfred NOBEL all died when
(named for Enrico Fermi)   (mendelevium)   (nobelium)

elements beyond uranium were still folkLORE.
(lawrencium)

75

Activity 8.5   Four new elements are named

Go to periodicvideos.com and watch the video about the naming of four new elements.

Activity 8.6   "Odd one out" (Which one of these doesn't belong?)

Now that we've finished our tour of the Periodic Kingdom, we can do an activity where you use your knowledge of the table to try to figure out which element doesn't belong in the group.

EXAMPLE 1:   Ne   Ar   Xe   Cl   (Chlorine does not belong because it is not a noble gas.)
EXAMPLE 2:   Sn   Bi   Ti   Al   (Titanium does not belong because it is not a true metal.)

1) Ca   Sc   Co   Ag

2) Rb   Be   K   Cs

3) S   C   K   P

4) Cd   Cm   Cr   Cu

5) Pm   Sm   Tm   Fm

6) Na   Ca   Ra   Ba

7) Pt   Pu   Pa   Np

8) B   Ge   Br   Si

9) Rh   Re   Rn   Ru

10) Te   Tb   Eu   Er

Which one of these doesn't belong?
K P H S

Hydrogen, because it doesn't belong to any group!

Activity 8.7   "Who am I?"

Figure out which lanthanide or actinide is being described.

1) I fluoresce bright green, so I am used in televisions and computer screens. _____

2) I was made by smashing a chromium atom into a lead atom. _____

3) I am the only radioactive lanthanide. _____

4) I am the last naturally occurring actinide. Elements above me are man-made. _____

5) If I were to lose an alpha particle, I would turn into uranium. _____

6) We are used to make very strong magnets. (The element right between us might also be magnetic, but it can't be used because it is dangerous.) _____ and _____

7) I was named after the man who invented the Periodic Table. _____

8) I can fluoresce either bright red or bright blue, depending on the atoms around me. _____

9) I was named after a mine in Sweden where I was first discovered. My name does not begin with Y, nor does it begin with T. _____

10) I was named after the US state where my discoverer lived and worked. _____

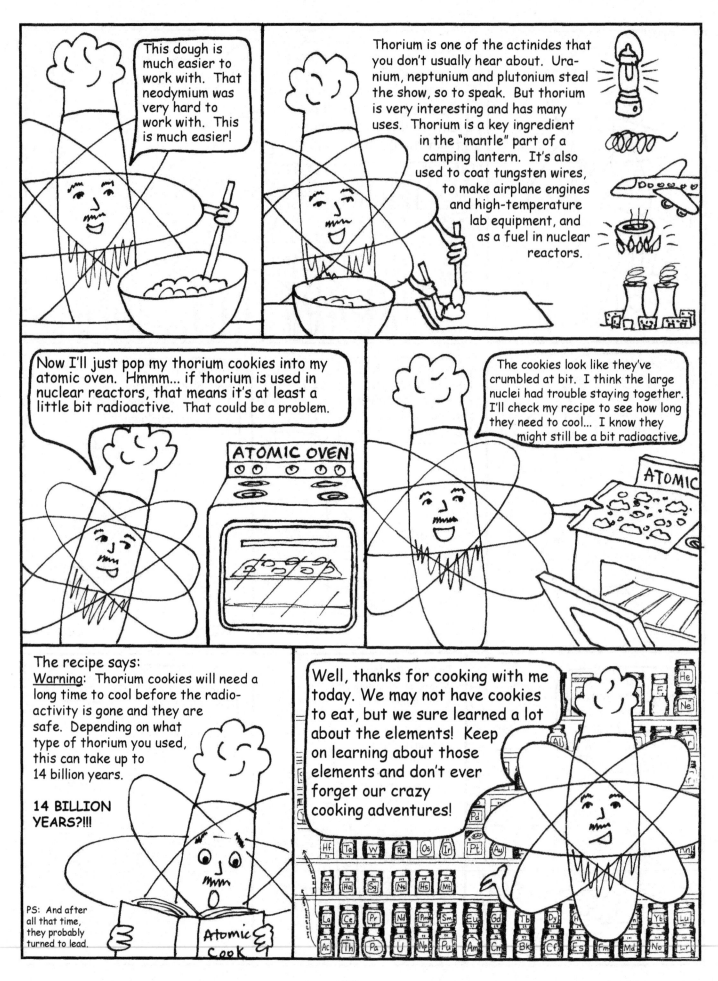

ANSWER KEY

ANSWER KEY

CHAPTER 1
Answers will vary for the activities not listed here.

Activity 1.5:
1) C= 2, O= 6   2) 3   3) 2   4) Si= 2, O= 8   5) 20

Activity 1.6:
1) nobelium   2) vanadium   3) gadolinium   4) polonium   5) einsteinium   6) berkelium
7) tellurium   8) scandium   9) ytterbium   10) niobium   11) tin   12) holmium
13) neptunium   14) curium   15) mercury   16) tantalum   17) cerium   18) gallium
19) selenium   20) bromine   21) iridium   22) thorium   23) nickel   24) cobalt   25) chlorine

CHAPTER 2
Activity 2.3
1) phosphorus   2) fluorine   3) calcium   4) gallium   5) titanium   6) silicon
7) rhodium   8) iodine   9) scandium   10) palladium   11) tin   12) sulfur
13) chlorine   14) argon   15) nitrogen   16) hydrogen   17) carbon   18) boron
19) potassium   20) xenon

Activity 2.4
1) "Why did the mouse say, "Cheep, cheep," when the bird's cage fell apart?"
"He was filling in for the bird who had the day off."
2) "What do you get when you cross a vampire with a mouse?"
"A terrified cat!"
3) What were Batman and Robin's new names after they were run over by a car?
Flatman and Ribbon!

CHAPTER 3
Activity 3.4
Nitrogen: $1s^2\ 2s^2\ 2p^3$
Sulfur: $1s^2\ 2s^2\ 2p^6\ 3s^2\ 3p^4$
Neon: $1s^2\ 2s^2\ 2p^6$
Chlorine: $1s^2\ 2s^2\ 2p^6\ 3s^2\ 3p^5$
Lithium: $1s^2\ 2s^1$
Boron: $1s^2\ 2s^2\ 2p^1$
Silicon: $1s^2\ 2s^2\ 2p^6\ 3s^2\ 3p^2$
Fluorine: $1s^2\ 2s^2\ 2p^5$

Activity 3.5
Ag-47	H-1	Os-76
Am-95	He-2	P-15
At-85	I-53	S-16
As-33	In-49	Se-34

Activity 3.6
1) Be   2) N   3) Na   4) S   5) P   6) Ca   Challenge: Fe

Activity 3.6
Just use the Periodic Table as your guide. It tells you what all the symbols are!

CHAPTER 4

Activity 4.1

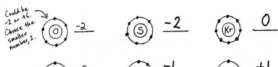

Could be -2 or +6. Choose the smaller number, 2. (arrow pointing to O) O: -2

S: -2   Kr: 0

Mg: +2   Br: -1   K: +1

C: ±4   P: -3   Ba: +2

Sc: +3   Si: ±4   Li: +1

Activity 4.2

B: +3   ·P: -3   ·Ö: -2   H· +1

·S̈: -2   Mg: +2   Äl: +3   Li· +1

:N̈e: 0   ·C̈: ±4   ·N̈: -3   :F̈l: -1

Li (lithium) +1   Li·

B (boron) +3   B̈:

I (iodine) -1   :Ï:

N (nitrogen) -3   ·N̈:

C (carbon) +4   ·C̈:

Be (beryllium) +2   Be:

K (potassium) +1   K·

S (sulfur) -2   ·S̈:

Activity 4.3

1) B- boron (boring)   2) Ar- argon ("are gone")   3) Ba- barium ("bury 'em")
4) Es- einsteinium (after Einstein)   5) Pu- plutonium (named after Pluto, which is way out there!)
6) Eu- europium (named after Europe)   7) Fe- iron (as in ironing clothes)   8) Kr- krypton
9) Hg - mercury   10) Cf- californium   11) Si- silicon ("silly con")   12) Ni- nickel
13) Po- polonium ("polo" like field hockey played while riding horses)   14) Os- osmium (sounds like "Oz")

CHAPTER 5 does not have any activities that need answers.

CHAPTER 6

Activity 6.4

1) PLaSTiC   2) GaSOLiNe   3) PrOPaNe   4) AsPHAlt   5) AlCoHoL   6) BUCKY BAIL
7) CaFFEINe   8) AmInO AcID 9) FAtS   10) PaRaFFIn   11) PoLYEsTeR   12) GLuCoSe
13) HaIr   14) GLuTeN   15) NYLON   16) ArTiFICIaI FLaVORS   17) PLaNT FIBeR
Trivia question: The letters J and Q do not appear in any element symbol, assuming you are using the most up-to-date version of the table. Uuq was a symbol until recently, so if you have an old table, you might see a Q.

Activity 6.5

1) b   2) d   3) a   4) noble gases   5) alkali   6) halogens   7) S   8) C   9) N   10) He
11) sand   12) salt   13) bleach   14) Teflon   15) plaster   16) c   17) e   18) protons
19) d   20) c   BONUS: 1) sulfur   2) 146   3) 38

CHAPTER 7

Activity 7.4

1) gallium, Ga   2) arsenic, As   3) lead, Pb   4) antimony, Sb   5) indium, In
6) tin, Sn   7) thallium, Tl   8) bismuth, Bi   9) tellurium, Te   10) aluminum, Al

CHAPTER 8

Activity 8.6

1) Ca (not transition metal)   2) Be (not alkali metal)   3) K (not non-metal)
4) Cm (not transition metal)   5) Fm (not lanthanide)   6) Na (not alkali earth metal)
7) Pt (not actinide)   8) Br (not semi-metal)   9) Rn (not transition metal)   10) Te (not lanthanide)

Activity 8.7

1) Tb, terbium   2) Sg, seaborgium   3) Pm, promethium   4) U, uranium   5) Pu, plutonium
6) Nd and Sm, neodymium and samarium   7) Md, mendelevium   8) Eu, europium
9) Er, erbium   10) Cf, californium

TEACHER'S SECTION

Reproducible patterns for
games and activities

Lab experiments

Group games

Skits

ACTIVITY IDEAS FOR CHAPTER 1

1) GROUP GAME: "Symbol Jars"

The purpose of this game is to learn the letter symbols of some of the common elements. Students do not need to have any previous knowledge. If players do already know some of the symbols, they can still play along with those who are beginners and just focus on the symbols that they don't know. More difficult symbols can always be added to the mix.

NOTE: This game is similar to the fishing game in chapter 2. If you are short on time, you might want to play just one or the other.

You will need: photocopies on card stock, scissors, pencils or crayons

Set up:

Photocopy the pattern page, (the empty bottles, page 92), onto white card stock. Make enough copies so that you have a bottle for each element you want to learn. Cut out the bottles. Write an element's symbol on the bottle, then write the name of the element on the back. (Important: Make sure you don't use anything that will bleed through to the other side! Also, don't press too hard, or your letters might show through to the back. This isn't the time to practice your engraving skills!) You could also reverse this, and put the name on the front and the symbol on the back. Either way is fine.

NOTE: If you are making several copies of the game so that you can play it with a class, make each set of cards a different color. If some cards get mixed up while the students are playing, they will be easy to sort back into their sets. (You won't end up with two of something in one set and none in another.)

How to play:

Before you begin, make sure each player has a little slip of paper with his name on it. Lay the jars out on the table in random fashion. Each player must "call" the jar he wants to play by saying the letter symbol. For example, a player might say, "C." Then the player has a choice: he can either "guess" or "peek."

If the player chooses "guess," he must say the name of the element that is represented by that symbol. After he says the name, he checks his answer by turning over the jar and reading the name on the back. If he is correct, he gets to pick up that jar and keep it. If not, he must leave the jar on the table.

If the player doesn't know a symbol and wants to learn it, he chooses the "peek" option. The player still begins by "calling" the jar he wants to play by saying the letter symbol. Then the player states his option, "peek," and turns the jar over to read the name on the back. After returning the jar to its original position, the player may then "reserve" the jar for his next turn by putting his slip of paper (with his name on it) on top of the bottle. No other player may call that jar while the name slip is on it. When that player's next turn comes around again, he can call that jar but this time use the "guess" option (assuming he does remember the name on the back—if he doesn't, he can always use the "peek" option again). If he guesses correctly, he keeps the jar.

The game is over when all the jars have been taken.

NOTE: If you are working from a paperback copy of this book, not a digital download, and you would like a digital file so that you can print these patterns using your computer's printer, go to www.ellenjmchenry.com, click on FREE DOWNLOADS, then on CHEMISTRY, and then you will see a link for "Printable pages for The Elements curriculum."

2) GROUP GAME: "Quick Six" (Round one—we'll play it again later with more cards!)

The purpose of this game is to become familiar with the names and numbers of the elements from hydrogen to xenon. Players do not need any previous knowledge for this game.

You will need: scissors, photocopies of the pattern pages (93-98) on white card stock, and colored pencils if you would like the students to color the cards (I suggest using the digital version of the curriculum to print the cards on your computer's printer, or get them printed at a print shop. If you need a digital file, see the note above, in italics.)

Set up:

Cut apart the cards. If you would like the students to add color to the cards, provide colored pencils and some extra coloring time.

<u>How to play:</u>

The object of the game is to be the first player to collect six cards.

Decide which player will be the "caller." This player must read from the list below instead of being one of the card players. If an adult is supervising the game, this is the obvious adult job. An adult caller may want to choose particular attributes from the list below to emphasize facts recently learned. It is easiest to go down the list in order, but the caller need not go in order, and may also use items from the list more than once. Feel free to add your own ideas to the list given below!

Each card player receives five cards, which he places face up in front of him. The rest of the cards go face down in a draw pile. The caller reads one of the attributes from the list (the first on the list if they are going in order). Each player looks at his five cards to see if he has a card that has that attribute. If he does, he slaps his hand down on the card. The caller looks to see who is the first player to slap his hand down. That player then shows the card under his hand. If the caller agrees that this card qualifies, then the player may remove that card from the line up and put it face down into a "keeper" pile. Then he draws a card from the draw pile to replace that card and restore him to five cards, face up.

NOTE: There's a chance that a student might know extra information about an element that is not on the card. If the adult in charge determines that the student's answer is accurate, I'd recommend allowing the student to use the information.

The caller then reads off another attribute from the list and the game continues in this manner until one player has six cards in his "keeper" pile. If no player has a card that qualifies, the caller simply goes on to the next one on the list.

NOTE: You might have to institute a rule that says only one slap per round. If they slap and get it wrong, the other players get to guess again, but they don't. Sometimes students slap before they read the card carefully. Using this rule will prevent careless slapping.

If you reach the end of the list below, just start over at the beginning again. (Or, better yet, add your own clues.)

A single game could take as little as 5-10 minutes, so play multiple games. You can switch callers between games.

Atomic number has a 3 in it
Name has two syllables
Used in lasers
Has something to do with the color green
Named after someplace in Scandinavia
Has something to do with teeth
Starts with the letter C
Atomic number has a 5 in it
Name has something to do with color
Used to make tools of some kind
Is named after a city (not a country)
Name has three syllables
Atomic mass does not contain a 0
Is used to make jewelry
Named after a country
Used for something that burns
Named after something in the solar system
Atomic number has a 7 in it
Is named after a country (not a city)
Used in fireworks
Has something to do with bones
Name starts with a vowel
Gemstones are made from it
Atomic mass is greater than 100
Is named after a country (not a city)
Name has a double letter, such as "dd" or "ss"
Symbol has only one letter
Atomic symbol contains a vowel
Is used to make some kind of medicine

Used in steel production
Used to repair the human body
Used in light bulbs
Is found as a gas in the air around us
Has something to do with eyes
Conducts electricity
Atomic mass is less than 50
Last three letters of the name are I-U-M
Name is from a Latin word
Is used in batteries or fuel
Has something to do with glass
First letter of name does not match first
letter of the symbol
Is found in some kind of gemstone
Name begins with the letter S
Name ends with letters O-N
Name starts with the "K" sound (C or K)
Is used in magnets of any kind
Used in something that makes light
Used to make coins
Symbol contains one of these letters: X, Y, or Z
Name has four syllables
Number has a 1 in it
Atomic mass contains a 1
Name begins with the letter R
Atomic number is a prime number
Is mixed into metal alloys

3) LAB DEMO: "A Recipe in Reverse" (Electrolysis of water)

In this experiment, you will start with H_2O and "break" it into its ingredients: H and O.

You will need a clear container, a 9V battery, a piece of cardboard (cereal box is fine), aluminum foil, tape, two pencil stubs (sharpened at both ends), water, salt

How to set it up:

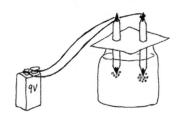

Think about how important the discovery of electricity was to chemists. If there was no way to separate water into its ingredients, how would you know it wasn't an element? Figuring out which substances were elements and which were not was a major puzzle for hundreds of years. Elecgtricity was necessary for the discovery of a number of elements including sodium, potassium and magnesium.

1) Put 2 teaspoons of salt into the cup of water and stir until dissolved.
2) Cut strips of aluminum foil and roll them into "wires." Curl one end around a battery terminal (tape in place if necessary) and put the other end around the sharpened pencil point and secure with tape. Make sure the graphite of the pencil is in good contact with the foil.
3) Push the pencils through the cardboard, as shown, so that the bottom points are in the water. (You can even strip off some of the wood with an X-acto knife if you want to, exposing more graphite. The more graphite showing, the more bubbles you will get.)

What will happen:

You will see bubbles forming around the ends of both pencils. If you look carefully, you will notice that there are about twice as many bubbles on one pencil as the other. Have the students guess which is which, by thinking about the recipe: H_2O. (Hint: The recipe says that for every oxygen atom there are two hydrogen atoms.)

4) SONG: "The Chemical Compounds Song" activity

Here is an activity you can do with this song. It can be done in pairs or in a group, whatever works in your situation. Use the song as the chant for any type of hand-slapping game, such as "Miss Mary Mack." ("Miss Mary Mack, Mack, Mack, all dressed in black, black, black...") The students may be able to suggest their favorite hand-slapping patterns. Any hand slap pattern will work, as long as both partners are doing the same pattern. For a large group, they can sit in a circle and slap thighs, then clap, then turn hands to the side and slap hands of person on their left and right simultaneously.

Also, there is a funny music video to watch on the YouTube playlist, made by a family who did this curriculum a few years ago. (www.YouTube.com/TheBasementWorkshop, click on "Show all playlists," then on "The Elements."

5) "MAKE FIVE" A game about mineral recipes

This game is recommended for older students, or those who are very enthusiastic about rocks and minerals. If "Symbol Jars" was enough, you can skip this game. You could also wait and play this game after the next chapter.

By definition, a mineral has a definite chemical composition (a recipe). In this game you will be introduced to the recipes for some common minerals. It's also an opportunity to keep on learning all those letter abreviations (symbols).

You will need: copies of the pattern pages copied onto card stock, scissors, and white glue (if you are assembling the paper dice) If you are using wooden cubes for the dice, you'll also need one or more markers.
(In a pinch for time, just take a fine point marker (red?) and write on real dice. Everyone can ignore the dots.)

NOTE: If you can get three wooden cubes, this is the best option. Most craft stores sell wooden cubes by the "each" or in small units and fairly inexpensively. If you want this game sturdy enough to survive future uses, consider using wooden cubes.

Preparation:
  1) Cut out the dice patterns (copied onto heavy card stock) and make into cubes, using small dabs of white glue on the tabs. (Or, write the symbols on wooden dice or even regular dice.)
  2) Cut apart the 16 mineral cards.

How to play:
Place the mineral cards on the table, face up, so they form a 4 x 4 square. Each player will have a turn rolling all three dice at once. The goal is to roll the ingredients to form a mineral. (One roll of the three dice per player per turn.) For example, if the first player rolls: Cu, Fe, and S, he should notice that those are the ingredients of chalcopyrite. Therefore, that player picks up the chalcopyrite card. If the next player rolls Ca, C, and WILD, he could make the wild card into O, and be eligible to pick up calcite.
The first player to collect five cards wins the game.

NOTE: If you are working from a paperback copy of this book, not a digital download, and you would like a digital file so that you can print these patterns using your computer's printer, go to www.ellenjmchenry.com, click on FREE DOWNLOADS, then on CHEMISTRY, and then you will see a link for "Printable pages for The Elements curriculum."

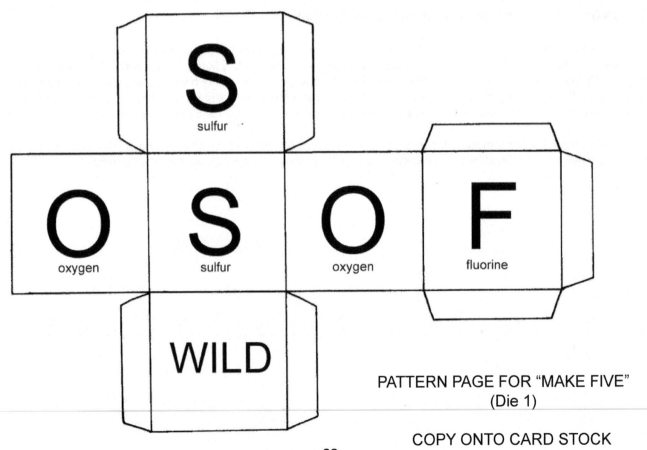

S
sulfur

O
oxygen

S
sulfur

O
oxygen

F
fluorine

WILD

PATTERN PAGE FOR "MAKE FIVE"
(Die 1)

COPY ONTO CARD STOCK

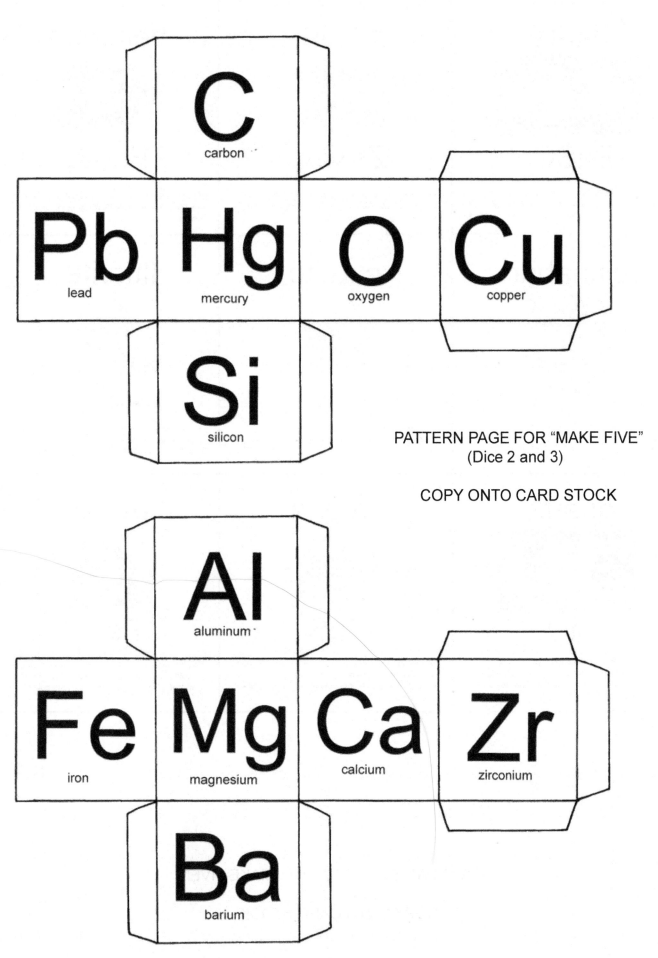

C
carbon

Pb
lead

Hg
mercury

O
oxygen

Cu
copper

Si
silicon

PATTERN PAGE FOR "MAKE FIVE"
(Dice 2 and 3)

COPY ONTO CARD STOCK

Al
aluminum

Fe
iron

Mg
magnesium

Ca
calcium

Zr
zirconium

Ba
barium

barite

$BaSO_4$

Often found in limestone or hot spring areas. Usually white or light brown. Sometimes crystalizes into rose shapes, which are popular with collectors.

zircon

$ZrSiO_4$

Found in nearly all igneous rocks, although in very small amounts. Because it is so hard, it is often used as a gemstone in jewelry.

hematite

Fe_2O_3

Hematite is a major ore (source) of iron. The name "hematite" comes from its blood-red color ("hema" means blood).

cinnabar

HgS

Cinnabar has a reddish color and is very dense (heavy) because of the mercury (Hg). Pure mercury is a liquid at room temperature, but it is a solid when bound to sulfur.

cuprite

Cu_2O

Cuprite forms cubic crystals. It is sometimes called "ruby copper" because of its color. When exposed to air it changes to CuO.

fluorite

CaF_2

Fluorite is used in the production of steel. It has a glassy luster and can look similar to a quartz crystal, except for its tetragonal (4-sided) shape.

quartz

SiO_2

Quartz is used in electronics, as a gemstone, and in the manufacturing of glass (where it is the main component). Sand is made of very tiny pieces of quartz.

galena

PbS

Galena is very dense (heavy) because of the lead in it. During the era of musket rifles, galena was used as the source of lead to make musket balls.

pyrite

FeS_2

This mineral is often called "fool's gold" because of its golden color and shiny luster. It has no actual gold in it. It leaves a black streak, not gold.

FIRST PATTERN PAGE FOR "MAKE FIVE"

COPY ONTO CARD STOCK

corundum

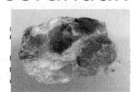

Al_2O_3

Corundum is very hard. It is so hard that it is used in industry as an abrasive (like sand paper). Blue corundum is called a sapphire and red is a ruby.

talc

$Mg_3Si_4O_{10}$

Talc is extremely soft. In fact, you can scratch it with your fingernail! Talc is the main ingredient in talcum powder (used to dry off after a shower).

calcite

$CaCO_3$

Calcite is the main ingredient in limestone. It is one of the most common minerals in the world. Caves are made of limestone.

gypsum

$CaSO_4$

Gypsum is a soft mineral. It is one of the main ingredients in plaster and plasterboard. One type of gypsum is called alabaster and was carved by ancient peoples.

chalcopyrite

$CuFeS_2$

Chalcopyrite is pinkish-purple with flecks of gold. It is found wherever copper is mined. The copper can be taken out of it by using chemical processes.

epsom salt

$MgSO_4$

This mineral dissolves into water very easily. It is often used in medical treatment of wounds on hands and feet. It helps in the healing process.

diamond/graphite

C

Strangely enough, both priceless diamonds and the stuff in your pencil are made of the same thing: pure carbon. The difference is how the atoms are bonded together.

SECOND PATTERN PAGE FOR "MAKE FIVE"

COPY ONTO CARD STOCK

PATTERN PAGE FOR "SYMBOL JARS"   COPY ONTO CARD STOCK

H 1
Hydrogen 1.0

Greek: "hydro–gen" (water-maker)

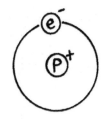

- Has no neutrons.
- Most abundant element in the Universe.
- Used in rocket fuel and fuel cells.

He 2
Helium 4.0

Greek: "helios" (sun)

- Used in balloons, blimps and scubing diving tanks.
- Discovered in the sun in 1895 using a spectrometer.

Li 3
Lithium 6.9

Greek: "lithos" (stone)

 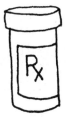

very small batteries

- Used in batteries, lubricants, medicines, and nuclear bombs.
- Is never found by itself in nature (it's always in a compound).

Be 4
Beryllium 9.0

from the mineral "beryl"

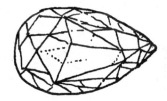

- Found in emeralds.
- Is mixed with copper to make "beryllium bronze," an alloy that will not create sparks.

B 5
Boron 10.8

from the compound "borax"

- Used to make heat-resistant glass.
- Used to make boric acid, which is used as an antiseptic eye wash.
- Used in nuclear power plants.

C 6
Carbon 12.0

Latin: "carbo" (charcoal)

- Diamonds, graphite and coal are all made of carbon.
- Carbon makes long chains (polymers) that are the basis of fossil fuels and plastics.
- Carbon is necessary for organic molecules found in living organisms.

N 7
Nitrogen 14.0

Greek: "nitron" (the mineral saltpetre)

- Most of the air we breathe is nitrogen.
- Used in air bags in cars.
- Doctors use liquid nitrogen to treat skin conditions.
- Proteins and DNA contain nitrogen.

O 8
Oxygen 15.9

Greek: "oxy-gen" (acid-maker)

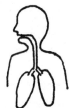

- Found in air, water and sand.
- Necessary for respiration and combustion.
- Ozone is made of pure oxygen.

F 9
Fluorine 18.9

Latin: "fluere" (to flow)

- Found in the mineral fluorite.
- Is put into toothpaste to fight cavities.
- Used as a coolant.
- Used in nuclear power plants.

"Quick Six" pattern page 1   93   Copy onto white card stock

Ne 10
Néon   20.1

Greek: "neo" (new)

- Used in neon lights and lasers.
- Neon never bonds to any other elements.

Na 11
Sodium   22.9

from soda ash

- Bonds with chlorine to make table salt.
- Used in street lights and in household cleaning products.
- Sodium is never found by itself in nature; it is always in a compound.

Mg 12
Magnesium   24.3

from Magnesia, in Greece

- Used in sparklers.
- Found in Epsom salts and "milk of Magnesia"
- Plants and animals need magnesium.

Al 13
Aluminum   26.9

from the compound "alumina"

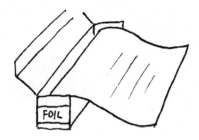

- Used in airplanes because it is so light and strong.
- Used for foil, tubes and cables.
- Used in fireworks.

Si 14
Silicon   28.0

Latin: "silex" (hard stone, boulder)

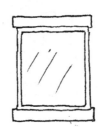

- Found in sand, clay, lava, glass and the mineral quartz.
- Used to make computer chips.

P 15
Phosphorus   30.9

Greek: "phosphoros" (bringer of light)

- Used in matches, fireworks, fertilizers and detergents.
- Discovered by an alchemist in 1669 as he was boiling down urine!

S 16
Sulfur   32.0

Latin: "sulfur" (stone that burns)

- Found in matches and fireworks.
- Used to vulcanize rubber.
- Volcanoes produce sulfur dioxide gas (a gas that's also produced by some factories and forms a large part of air pollution).

Cl 17
Chlorine   35.4

Greek: "kloros" (light green)

- Bonds with sodium to make table salt.
- Used to disinfect swimming pools.
- Is an ingredient in PVC plastics.
- Combines with hyrdogen to make HCl, an acid that your stomach produces to help with digestion.

Ar 18
Argon   39.9

Greek: "argos" (lazy)

- Used in lightbulbs and lasers.
- Does not bond to, or react with, any other element.

K 19
Potassium 39.0

from the word "potash"

- Used in fertilizers.
- Is an ingredient in gun powder.
- Bananas contain a lot of potassium.
- Can form salts, just like sodium can.

Ca 20
Calcium 40.0

Latin: "calx" (chalk)

 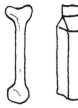

- Found in chalk, limestone, plaster, concrete, bones, and teeth.
- Milk contains a lot of calcium.
- Calcium in water makes it "hard."

Sc 21
Scandium 44.9

named after Scandinavia

- Used in stadium lighting.
- Used in large television screens.
- Radioactive scandium is used as a "tracer" in petroleum refineries.

Ti 22
Titanium 47.9

named after the Greek Titan gods

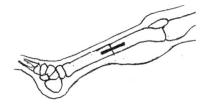

- Used to repair bones.
- Because it is lightweight it is used in airplane motors.
- Is an ingredient in paint pigments.

V 23
Vanadium 50.9

after the Scandinavian goddess Vanadis

 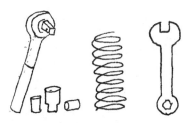

- Used in making steel.
- Is an ingredient in metals that are used to make tools, springs and engines.

Cr 24
Chromium 51.9

Greek: "chroma" (color)

- Gives rubies their red color.
- Used to make red, green and yellow paint.
- Used as a shiny coating for metals.
- Used to make video tapes.

Mn 25
Manganese 54.9

Latin: "magnes" (magnetic)

- Added to steel that needs to be very strong (for example: rifle barrels and bank vaults).
- Is necessary for the functioning of vitamin B1 in our bodies.

Fe 26
Iron 55.8

from Old English "iren"

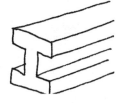

- Discovered in ancient times.
- Used in steel and in magnets.
- Found in red blood cells and in rust.
- Meteorites often contain iron.
- Red rocks usually contain iron.

Co 27
Cobalt 58.9

German "kobald" (evil gnomes)

- Miners used to say "kobald" lurked in the mines (and the name stuck).
- Used in "alnico" magnets.
- Used in making drill bits and razors.
- Can be used to color glass deep blue.

Ni 28
Nickel 58.7

German: "Nickel" (Satan)

- Name comes from "Kupfernickel," meaning "Satan's copper."
- Used in the coloring of glass.
- Used to make coins and utensils.

Cu 29
Copper 63.5

Latin: "Cuprum" (from Cyprus)

- Used for coins, wires and pipes.
- The Statue of Liberty is made of copper.
- Copper mixed with zinc makes brass.
- Copper mixed with tin makes bronze.

Zn 30
Zinc 65.4

Greek: "zink"

- Used for galvanizing (protecting) metals such as iron and steel.
- Zinc sulfide glows in the dark.
- Zinc oxide is used in photocopiers.

Ga 31
Gallium 69.7

Latin: "Gallia" (France)

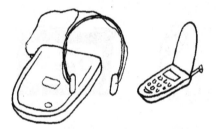

- Gallium arsenide is used in lasers and in compact disc players.
- Used in cell phones and in medical devices.

Ge 32
Germanium 72.6

Latin: "Germania" (Germany)

semi-conductor

lens

- Is a semi-conductor and therefore is used in transitors.
- Used in lenses and fiberoptics.

As 33
Arsenic 74.9

Latin: "arsenicum" (a pigment)

- Famous for its use as a poison.
- Is an ingredient in weed killers and insecticides.
- Used in lasers and LED's.

Se 34
Selenium 78.9

Greek: "selene" (moon)

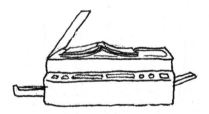

- Used in photocopiers because it conducts electricity in the presence of light.
- Used in robotics and in light meters.
- Selenium is beneficial to our bodies and acts as an anti-oxidant, protecting use from cellular damage.

Br 35
Bromine 79.9

Greek: "bromos" (stench)

- Bromine is a reddish liquid with a very bad smell.
- Found in sea water and salt mines.
- Used in photographic film.

Kr 36
Krypton 83.8

Greek: "kryptos" (hidden)

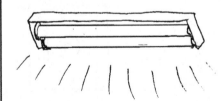

- Used in fluorescent flight, especially photographic bulbs.
- Used in UV lasers and in atomic clocks.

Rb 37
Rubidium 85.5

Latin: "rubidus" (deep red)

Rubidium captures atoms of gases that should not be in a vacuum jar or tube..

- Is a by-product of the refinement of lithium and cesium.
- Used as a gas "scavenger" (collector) in vacuum tubes.

Sr 38
Strontium 87.6

after the Scottish village of Strontia

- Used in fireworks (bright red).
- Used in batteries in ocean buoys.
- Used to produce beta radiation.
- Used to research bone structure.

Y 39
Yttrium 88.9

after the Swedish town of Ytterby

a moon rock

- Used in superconductors and lasers.
- Rocks from the moon contain yttrium.
- Used to make the bright red color in television screens.

Zr 40
Zirconium 91.2

Arabic: "zargun" (gold color)

- Made into gemstones.
- Used in catalytic converters in cars.
- Used for heat-resistant parts in nuclear power plants and in space shuttles.

Nb 41
Niobium 92.9

named after the Greek goddess Niobe

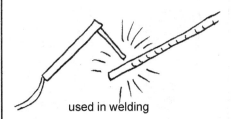

used in welding

- Used in welding rods, cutting tools, and superconducting magnets.
- Is added to steel to make it heat-resistant.

Mo 42
Molybdenum 95.9

Greek: "molybdos" (lead)

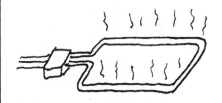

- Used for filaments in heaters.
- is an ingredient in steel that is used to make engines for cars and planes.
- Large deposits of molybdenum are found in Colorado.

Tc 43
Technitium 99.0

Greek: "teknetos" (artificial)

- Is radioactive.
- Not found in nature. Must be made in a nucluear laboratory.
- Is combined with other elements and used in medical procedures.

Ru 44
Ruthenium 101.1

Latin: "Ruthenia" (Russia)

 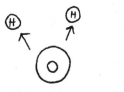

- Used to split water molecules.
- Used in the jewelry making industry.
- Often mixed with titanium and platinum to increase their hardness.

Rh 45
Rhodium 102.9

Greek: "rhodon" (rose)

electrode

- Rhodium salts have a rose color.
- Used in catalytic converters in cars.
- Used in headlight reflectors.
- Used in jewelry to prevent tarnishing of sterling silver.
- Combined with **Pt** and **Pd** to make spark plugs, electrodes, and other electronic parts.

Pd 46
Palladium   106.4

named after the asteroid Pallas

- Used in dentistry and in jewelry.
- Used in catalytic converters in cars.
- Used to purify hydrogen gas.
- Used for treatment of tumors.

Ag 47
Silver   107.8

Anglo-Saxon: "soilful" (silver)
Symbol from Latin "argentum"

- Used to make coins, jewelry, mirrors, silverware, photographic film and electronic components.
- Sterling silver contains copper.

Cd 48
Cadmium   112.4

Greek: "kadmeia" (earth)

- Used in rechargeable batteries.
- Is a neutron-absorber in nuclear reactors.
- Used to make yellow and red pigments in paints.

In 49
Indium   114.8

Latin: "indicum" (indigo blue)

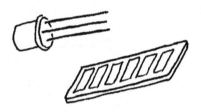

- Used in transistors and solar cells.
- Often mixed with other metals to make alloys.
- Its light wave pattern in a spectrometer shows bright purple lines.

Sn 50
Tin   118.7

Latin: "stannum" (tin)

pewter   bronze

- Is an ingredient of pewter.
- Is mixed with copper to make bronze.
- Turns into powder at low temperatures.

Sb 51
Antimony   121.7

Greek: "anti-monos" (not alone)
Symbol comes from "stibnium"

GLAZE   BATTERY   SOLDER

- Is also known by the name Stibnium.
- Used in ceramics, glazes, solder, lead batteries and matches.
- Increases hardness in alloys.

Te 52
Tellurium   127.6

Latin: "tellus" (earth)

rubber

- Used to "vulcanize" rubber (although sulfur is the key ingredient in vulcanization)
- Is one of the few elements that will bond with gold.
- Used to color glass.
- Used in ceramics.

I 53
Iodine   126.9

Greek: "iodes" (violet)

- Used as a disinfectant.
- Used in halogen lamps, ink pigments and photographic film.
- Our thyroid glands need iodine.

Xe 54
Xenon   131.3

Greek: "xenos" (strange)

- Used in camera flash bulbs, strobe lights, UV lamps and tanning bed lamps.

ACTIVITY IDEAS FOR CHAPTER 2

1) GROUP GAME: Play the element fishing game

<u>Overview of game</u>: This game can be played with multiple skill and age levels. The rules are constructed so that students who have no previous knowledge of the elements can play with those who do. The rules are similar to the "Symbol Jars" game from chapter 1, so if you played that game then this one will not require a lot more initial instruction. (The kids won't mind the repetition. I've never had any student complain about playing both jars and fish.)

<u>You will need</u>: copies of the fish pattern page, printed onto heavy card stock if possible. You may use as few or as many of the elements in your game as you wish. Just copy enough fish for the number of elements you want to use.

<u>You will also need</u>: scissors, pencils or crayons, string, paper clips, little slips of paper (one per player), at least one pole of some kind (you can make as many fishing rods as you want to), at least one magnet (one magnet per pole), an area marked off to be the "pond"
     TIP: When I've played this with a large group, it turns into total chaos if there are too many fishing poles. If the kids have a rod in their hands, they WILL fish whether it is their turn or not. I recommend small groups (no more than 4-5 kids per "pond") with only one rod to pass around, two at most.
     OPTIONAL: Have students trade in their paper fish for edible fish crackers at the end of the game.

<u>Set-up</u>:
  1) Cut out the paper fish. Have the students write the name of an element on one side of the fish and its symbol on the reverse side. (NOTE: Make sure you use a writing implement that does not bleed through the paper.) If you have limited class time, you may want to have the fish pre-labeled before class. If you want to make your fish durable, you could laminate them.
  2) Put a paper clip on the nose of each fish. Make a fishing pole from a rod (even a yardstick will do) and a string, and put a magnet on the end of the string.
  3) Mark off an area that will be the fishing pond. If you want to get fancy, you can use a plastic wading pool. (I've used blue painter's tape on both hard floors and carpets.)
  4) Each player needs a slip of paper with his or her name on it.
  5) After the fish are made, put them into the pond so that either all the names or all the symbols are facing up. The game seems to be easier to play if the symbols are facing up. We've found that reading the name and guessing the symbol is a little more difficult.

<u>How to play</u>:
     The rules are very similar to "Symbol Jars." The swinging strings add a new dimension to the game, in that you don't always bring up the fish that you called out. Oh well, it's part of the game!
     Each player must "call" the fish before he puts his rod in, by saying the name or letter on it. He must also choose one of two options: "guess" or "peek"   If he chooses "guess" this means that he will try to say what is on the reverse side before he pulls the fish out of the pond. After guessing, he reels in the fish and looks on the back. If he is right, he keeps the fish; if not, the fish goes back in the water. The other option, "peek," is for when the player has no idea of the right answer and needs to learn it. The player still "calls" the fish, but then says "peek." After reeling in the fish, the player reads the reverse side out loud. After the "peek" option, the player may then put his name slip under the paper clip, before returning the fish to the pond. This reserves the fish for him until his next turn. No other player may catch his fish in the interim. On his next turn, that player will probably want to use the "guess" option, remembering what he read on the back of the fish last time. If he remembers correctly, he keeps the fish. If not, the fish goes back in the pond (with no name slip).

     The game is over when all the fish are gone. You could play for a winner by counting up who has the most fish, or you can make the game non-competitive and simply give the players an edible reward for each fish they caught.

Even my middle school students enjoy fishing.

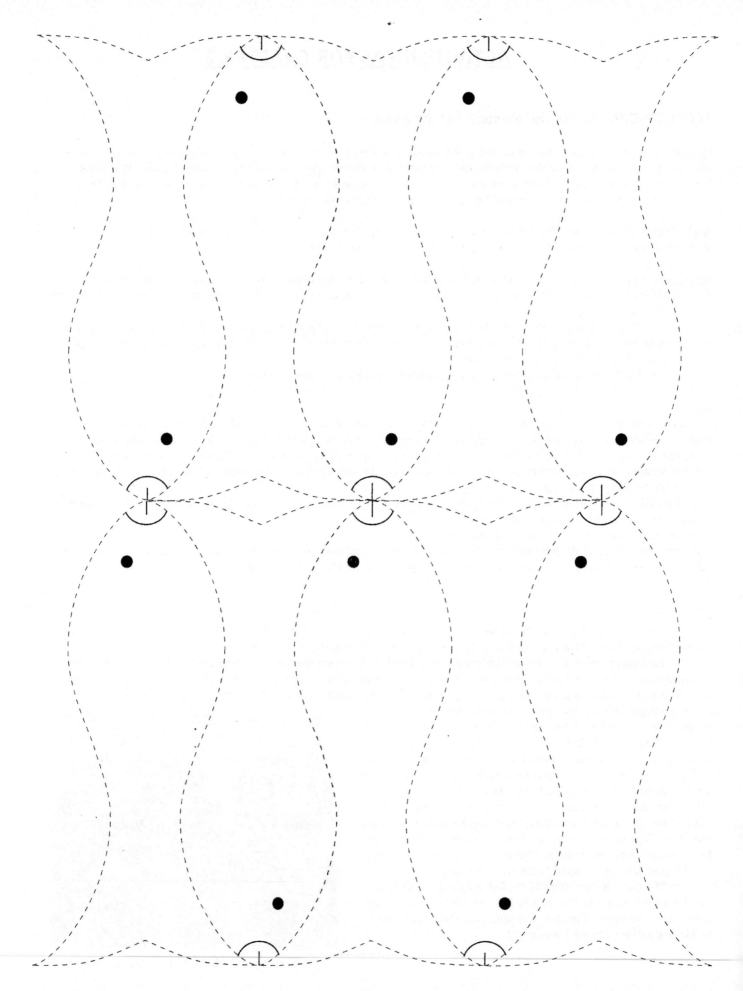

100

2) ACTIVE GROUP GAME: The Periodic Table Jump Rope Rhyme

You will need: jump ropes and the audio track (www.ellenjmchenry.com/audio-tracks-for-the-elements)

      Note: You can have the students use individual ropes, or you can do it as a group activity with one long rope and a "turner" at each end. This second method is nice to start with because the turners and the players who are not jumping can be the ones to recite the elements while the jumper concentrates on jumping. I have found that it is not hard at all to elicit very loud group chanting as players jump. It's kind of natural to join in with a group chant. So the students spend a lot of time chanting the rhyme over and over again. They can't help but remember at least some of it eventually.

      You might want to start by just listening to the audio track to catch on to how the chant goes. After a few times you can dispense with the audio track and have the students do their own chanting.   The game will be to see who can jump all the way to krypton without missing the rope. A player who misses the rope has to start back at hydrogen again (which is GREAT because that makes everyone review!). You might want to allow kids who miss before beryllium to get another try. NOTE: If you have kids who are shy about doing "overhead" jumping, you can do what we called "swayzees" (way back in the 1970s and 80s). Just swing the rope back and forth, never letting it go higher than waist high.

3) MATCHING CARD GAME: A way to review the info from the Chemical Compounds Song

You will need: picture cards showing the items named in the song (water, salt, bleach, rust, etc.) and cards with the chemical recipes written on them. Cards should all be the same size and made out of card stock, if possible, so that the writing does not show through on the back.

      NOTE: Apologies that the pictures can't be provided, but there could be copyright issues involved. However, they are easily found in seconds with Google image search. Just take a few minutes and clip them off Google and print them however your computer is set up to do such a task. I put 6 pictures on a page, but you can do more or less.

      ALSO NOTE: If you want to make the card game a little harder for older students, add these chemical formulas: H_2O_2 (hydrogen peroxide), N_2O (nitrous oxide or laughing gas), H_2S (hydrogen sulfide, the smell of rotten eggs), and NH_3 (ammonia, the smell of glass cleaning products or wet diapers).

How to play: Use standard "Memory match" game rules.

ACTIVITY IDEAS FOR CHAPTER 3

1) ACTIVE GROUP GAME: Human model of atom (a good outdoor activity)

You will need: as many players as possible, and a large outdoor (or indoor) space.

How to play:
1) All players will represent electrons. The adult (or a stationary object) will represent the nucleus. Assign each player either a number (starting from one and going up as high as you have players). This number corresponds to an element. 1=hydrogen, 2-helium, etc.
2) Explain that this model will be a "solar system" model, with the electrons going in circles around the nucleus.
3) As the number/name of each element is called out, the player who has been assigned that element comes to join the atom. The first two players must run, without bumping into each other, in a fairly tight circle, not too far from the nucleus. The third player, when called, must start a new circle, farther out than the first one. Then players are added, one at a time, until the second ring has eight in it. If you have more players, start a third ring.
4) While players are being added, the ones already in the atom must keep going.
5) For an added learning bonus, tell your players that you'll count to ten and in that time they must circle the nucleus one time. Just one time. They will find, of course, that the players in the outer ring will have to run very fast in comparison with the ones close to the nucleus. This analogy can be helpful when trying to explain how electrons can be "high energy" (a concept you meet in photosynthesis, cellular respiration). The outer shell electrons have more energy than the inner ones. If an electron gets zapped by a photon of energy it can jump to a higher shell. It can't stay there, though and when it comes back down it must release that energy again, and sometimes that energy is light we can see. (This is how glow-in-the-dark pigments work.)
    NOTE: While adding more players, keep an eye on the ones already running and make sure they stay at opposite ends of the circle and don't run into each other. Electrons never, ever, ever run into each other! They like to be in pairs, but at the same time, they like to be on opposite sides.

2) ACTIVITY: The Quick and Easy Atomizer

This can be used as a group activity. The directions are in the student booklet.

3) ACTIVE GROUP GAME: Jump Rope Rhyme Challenge, Round 2

Do the jump rope challenge again, only this time let the students go as far as they can on the Periodic Table. The jumper will not be the one reciting. The players who are not jumping are responsible for reciting. Use the rhyme up to krypton, then start saying the elements in order after that.

You may want to go over pronunciation ahead of time. You can use a dictionary, but there is also a helpful pronunciation guide at the beginning of this book. Also if you type "how to pronouce [your word]" into YouTube, you will be offered several pronunciation guides.

4) ANOTHER ONLINE INTERACTIVE GAME

Here is another online game you can use for more Periodic Table practice. It features a cartoon mouse on the home-page, Proton Don. (The mouse doesn't seem to be part of the game, though.) This game is a way to keep learning all those letter symbols. http://www.funbrain.com/periodic/

ACTIVITY IDEAS FOR CHAPTER 4

1) TABLE GAME: The Periodic Table Game

You will need: copies of the four pattern pages, assembled to make the Periodic Table, coins (about 5 pennies, 5 nickels, 5 dimes, and 1 quarter per player), a tuna can or small plastic container of similar size, a pair of standard dice, some tokens (one per player, and they don't have to be real game tokens, you can use anything), and some black paper squares the size of one space on the game board (two rectangles per player)

You may also want to make copies of the list of names and places for the students to study BEFORE the game starts. Once the game starts, no peeking at the list. Of course, the list may have to be consulted during the game to check answers.

NOTE: Again, if you need a digital file to print from, you can find one at www.ellenjmchenry.com in the chemistry section of the free downloads.

About the game board:

The number in the upper right hand corner of each square is the valence number. It is the number of electrons the element would like to receive or give away. Many elements (especially in the middle of the table) have more than one valence number. I've chosen just to list the highest valence for each element. It simplifies the game considerably and makes the mathematical pattern of the table more obvious. However, you may want to make your players aware that in reality many of the elements can have more than one valence number. In this game, the elements in each column end up displaying the same valence, which is a basic concept in learning to understand the Periodic Table. The word "periodic" means it has repeating patterns, and the valencies are one of these patterns. Notice that the last five elements do not have a valence number listed. These elements only exist for a fraction of a second and therefore their valence cannot be determined.

The large letters in each box are the letter symbols for each element.   Underneath the letter symbol is the name of the element.

Most elements are solids at room temperature. Notice that the elements that are liquids at room temperature are marked with a liquid drop, and those that are gases at room temperature are marked with a gas cloud.

There is a strange break at two places in the Periodic Table. One is after Lanthanum and one is after Actinium. These extra sections are listed at the bottom of the table simply because inserting them in the middle of the table would make the table too wide to fit comfortably on a page. There's no scientific reason for putting them at the bottom_it's simply a graphics decision.

The black and white version of the game is identical to the colorized version (except for the color, of course). The black and white version is for students who love to color, or who will learn more by coloring their own table. The black and white version is also cheaper and easier to reproduce.

How to play:

Before starting the game, players get a chance to study the information page that lists elements named after people and places. You might want to make additional photocopies of it. Once the game starts, no peeking except to check answers.

1) Put all the coins in the can and place it on the circle marked BANK. Put the players' tokens on START. Give each player 5 pennies to begin with.
2) Players take turns moving the number of spaces they roll on the dice. (Use two dice so the game doesn't go too slowly.) Unless your tokens are pretty small, you will probably want to allow only one token per square. Players will have to jump over each other. It's up to you whether to count that hop over another player as one of your actually "hops" or or not. Either way is fine as long as everyone agrees to the rules ahead of time and abides by them while playing.
3) When a player lands on a space, he looks at the valence number, which is in the upper right corner. If it is a positive number, he takes that many pennies from the bank. If the number is negative, he loses that many pennies and must put them into the bank.
4) Certain elements have special features:
   GASEOUS ELEMENTS (indicated by a cloud shape): extra roll
   LIQUID ELEMENTS (indicated by a droplet shape): extra roll
   PRECIOUS METAL: bonus of three pennies (Precious metals include silver, gold, platinum. You may add others to your list if you want to, as long as everyone agrees.)
   RADIOACTIVE ELEMENTS: The radioactive elements have little "shine" lines around their letter symbols. The player must place a square black shield on the spaces before and after that space, to keep other players "safe." No one can land on a black shield. If other players come past while the shields are in place, they simply hop over all three spaces (the two with the black shields and the one in the middle that has a token sitting on it) and keep going with their turn. Those three spaces do not count at all (they do not use up three hops). Just ignore those three spaces as if they were not there.   When it is the radioactive player's turn again, he removes the black shields and simply proceeds with his turn.
   ELEMENT NAMED AFTER A PERSON OR PLACE: If a player lands on an element that he thinks was named after a person or a place, he may take a 3 penny bonus if he can name that person or place. If he is wrong, he does not get the bonus, but there is no penalty for guessing.
   LANTHANIDES and ACTINIDES: Don't forget about these rows! After a player lands on lanthanum, he goes down to the lanthanide series. At the end of the row, he hops back up to hafnium. Similarly, after actinium comes thorium. After that row, hop back up to the main table and continue on with rutherfordium. (Often players forget the lanthanides the first time they play the game. If this happens and it's discovered too late to go back, you may want to just have the other players skip the lanthanides also, to make it fair play for everyone.) We don't know much about these rows yet, and they may seem like an annoyance in the game, but we'll find out in chapter 8 how incredibly important some of these are to our modern lifestyle (computers, cell phones, ipads, etc.).
5) At any time during the game a player may "make change," trading in pennies for nickels or dimes. The bank needs to have a good supply of pennies all the time, so when that supply gets low, players must make change to restock the bank.
6) You do not have to land on FINISH with an exact roll. After all players reach FINISH, the game is over. The player with the most money wins. (But everyone wins if you all learn and have fun!)

ELEMENTS NAMED AFTER PLACES:

Americium: America
Berkelium: Berkeley, CA
Californium: California
Cerium: the asteroid Ceres
Erbium: Swedish town of Ytterby
Europium: Europe
Francium: France
Gallium: France (Gall was the ancient name for France)
Germanium: Germany
Hafnium: Hafnia is Latin for Copenhagen, Denmark
Holmium: Stockholm, Sweden
Neptunium: the planet Neptune
Palladium: the asteroid Pallas
Plutonium: the until-recently-a-planet Pluto
Polonium: Poland
Rhenium: the Rhine area of Germany
Ruthenium: the province of Ruthenia in the Czech Republic
Scandium: Scandinavia
Strontium: Scottish town of Strontian
Tellurium: the planet Earth (the Greek word is Tellus)
Terbium: the Swedish town of Ytterby
Thulium: Scandinavia (the ancient name for Scandinavia was Thule)
Uranium: the planet Uranus
Ytterbium: the Swedish town of Ytterby
Yttrium: again, the Swedish town of Ytterby

ELEMENTS NAMED AFTER PEOPLE:

Curium: Marie Curie, discoverer of radium and polonium
Einsteinium: Albert Einstein
Fermium: Enrico Fermi, a physicist during the World War II era
Gadolinium: Johan Gadolin, a Finnish chemist
Gallium: Lecoq de Boisbaudran, a 19th century chemist (Gallus is Latin for "cock")
Lawrencium: Ernest O. Lawrence, a 20th century physicist
Mendelevium: Dmitri Mendeleyev, inventor of the Periodic Table
Meitnerium: Lise Meitner, a 20th century physicist
Mercury: Mercury, mythological Roman god
Niobium: Niobe, the daughter of mythological Greek god Tantalus
Nobelium: Alfred Nobel, inventor of dynamite, and namesake of the Nobel Prize
Niels-Bohrium: Niels Bohr, a 20th century chemist and physicist
Promethium: Prometheus, mythological Greek god who gave fire to mankind
Seaborgium: Glenn Seaborg, a 20th century chemist and physicist
Tantalum: Tantalus, mythological Greek god
Tin: Tinia, mythological Etruscan god ("Sn" comes from its Latin name, stannum)
Thorium: Thor, mythological Norse god of thunder
Vanadium: Vanadis, mythological Scandinavian goddess

Start

THE Periodic Table Game

Liquid or Gas at Room Temperature > **Roll Again**

Radioactive > **Put up Shields on Either Side**

Named After Person or Place > **3 Extra Pennies if You Name It!**

| ▨ | ☒ | ☐ |

1 +1									
H Hydrogen									

3 +1	4 +2								
Li Lithium	**Be** Beryllium								

11 +1	12 +2								
Na Sodium	**Mg** Magnesium								

19 +1	20 +2	21 +3	22 +4	23 +5	24 +6	25 +7	26 +3	27 +3
K Potassium	**Ca** Calcium	**Sc** Scandium	**Ti** Titanium	**V** Vanadium	**Cr** Chromium	**Mn** Manganese	**Fe** Iron	**Co** Cobalt

37 +1	38 +2	39 +3	40 +4	41 +5	42 +6	43 +7	44 +3	45 +3
Rb Rubidium	**Sr** Strontium	**Y** Yttrium	**Zr** Zirconium	**Nb** Niobium	**Mo** Molybdenum	**Tc** Technetium	**Ru** Ruthenium	**Rh** Rhodium

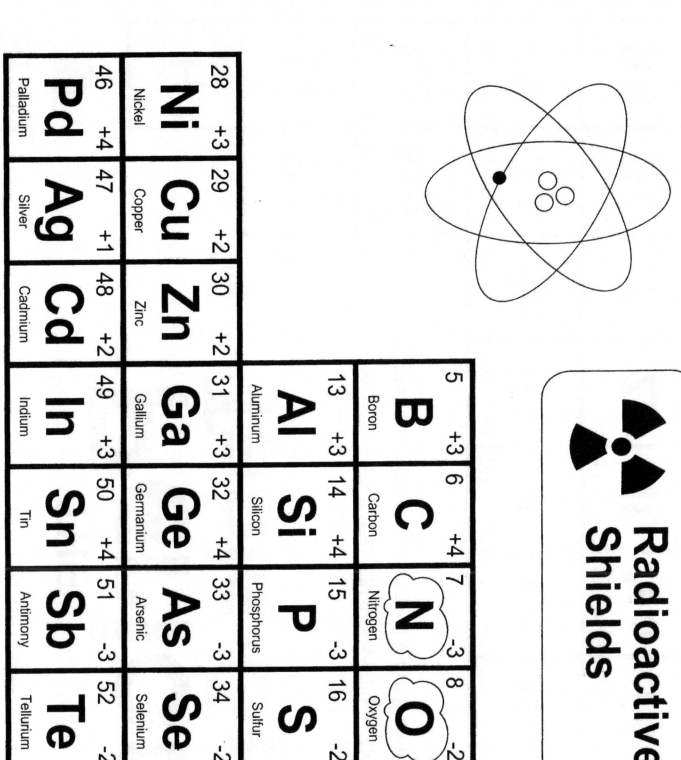

Radioactive Shields

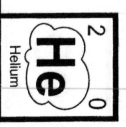

28 Ni +3 Nickel	29 Cu +2 Copper	30 Zn +2 Zinc					
46 Pd +4 Palladium	47 Ag +1 Silver	48 Cd +2 Cadmium					

5 B +3 Boron	6 C +4 Carbon	7 N -3 Nitrogen	8 O -2 Oxygen	9 F -1 Fluorine	10 Ne 0 Neon	2 He 0 Helium
13 Al +3 Aluminum	14 Si +4 Silicon	15 P -3 Phosphorus	16 S -2 Sulfur	17 Cl -1 Chlorine	18 Ar 0 Argon	
31 Ga +3 Gallium	32 Ge +4 Germanium	33 As -3 Arsenic	34 Se -2 Selenium	35 Br -1 Bromine	36 Kr 0 Krypton	
49 In +3 Indium	50 Sn +4 Tin	51 Sb -3 Antimony	52 Te -2 Tellurium	53 I -1 Iodine	54 Xe 0 Xenon	

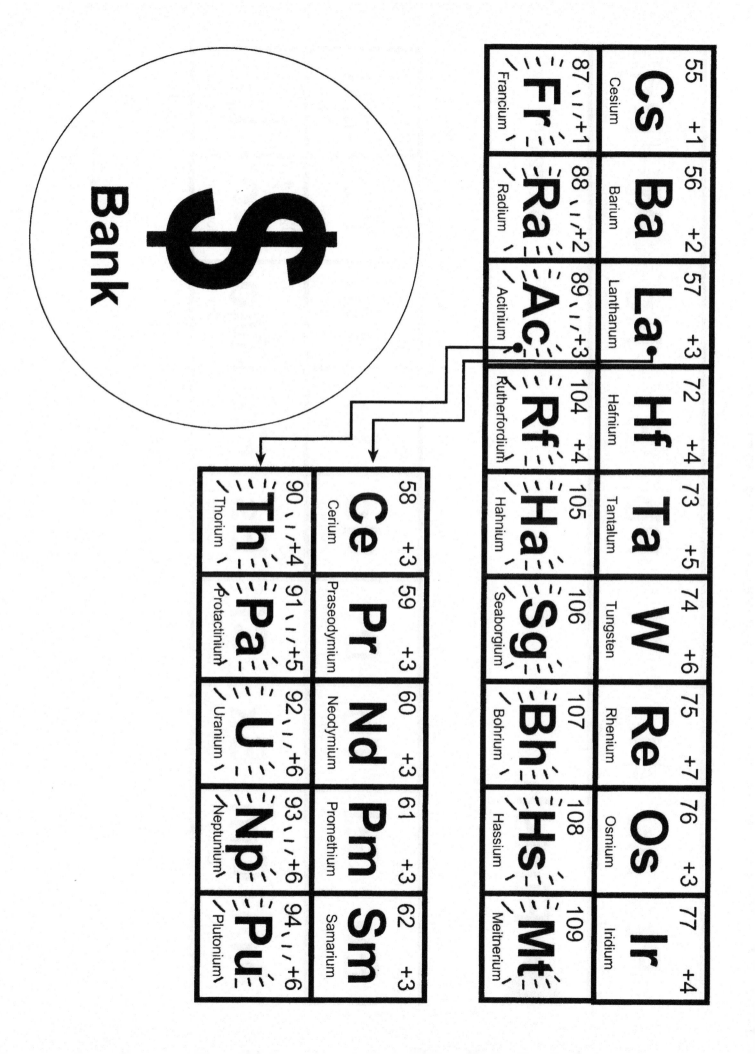

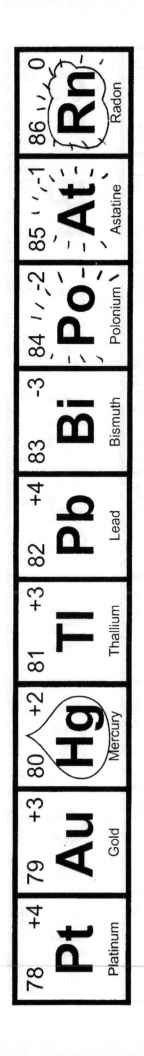

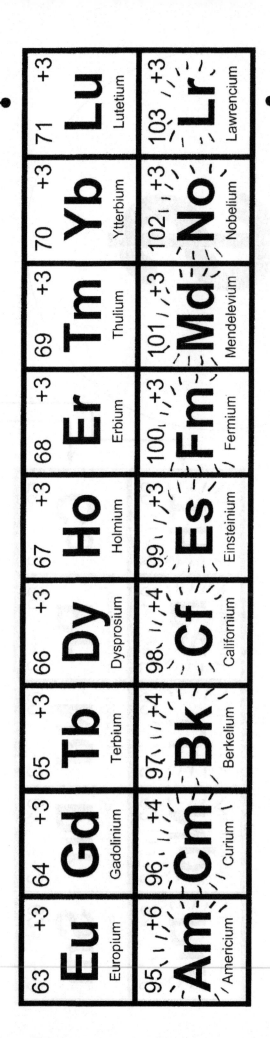

78 +4	79 +3	80 +2	81 +3	82 +4	83 -3	84 -2	85 -1	86 0
Pt	**Au**	**Hg**	**Tl**	**Pb**	**Bi**	**Po**	**At**	**Rn**
Platinum	Gold	Mercury	Thallium	Lead	Bismuth	Polonium	Astatine	Radon

▲ Finish

Go to 72 (Hf)

63 +3	64 +3	65 +3	66 +3	67 +3	68 +3	69 +3	70 +3	71 +3
Eu	**Gd**	**Tb**	**Dy**	**Ho**	**Er**	**Tm**	**Yb**	**Lu**
Europium	Gadolinium	Terbium	Dysprosium	Holmium	Erbium	Thulium	Ytterbium	Lutetium
95 +6	96 +4	97 +4	98 +4	99 +3	100 +3	101 +3	102 +3	103 +3
Am	**Cm**	**Bk**	**Cf**	**Es**	**Fm**	**Md**	**No**	**Lr**
Americium	Curium	Berkelium	Californium	Einsteinium	Fermium	Mendelevium	Nobelium	Lawrencium

Go to 104 (Rf)

THE Periodic Table Game

Liquid or Gas at Room Temperature > **Roll Again**

Radioactive > **Put up Shields on Either Side**

Named After Person or Place > **3 Extra Pennies if You Name It!**

Start

1 H +1 Hydrogen								
3 Li +1 Lithium	4 Be +2 Beryllium							
11 Na +1 Sodium	12 Mg +2 Magnesium							
19 K +1 Potassium	20 Ca +2 Calcium	21 Sc +3 Scandium	22 Ti +4 Titanium	23 V +5 Vanadium	24 Cr +6 Chromium	25 Mn +7 Manganese	26 Fe +3 Iron	27 Co +3 Cobalt
37 Rb +1 Rubidium	38 Sr +2 Strontium	39 Y +3 Yttrium	40 Zr +4 Zirconium	41 Nb +5 Niobium	42 Mo +6 Molybdenum	43 Tc +7 Technetium	44 Ru +3 Ruthenium	45 Rh +3 Rhodium

Radioactive Shields

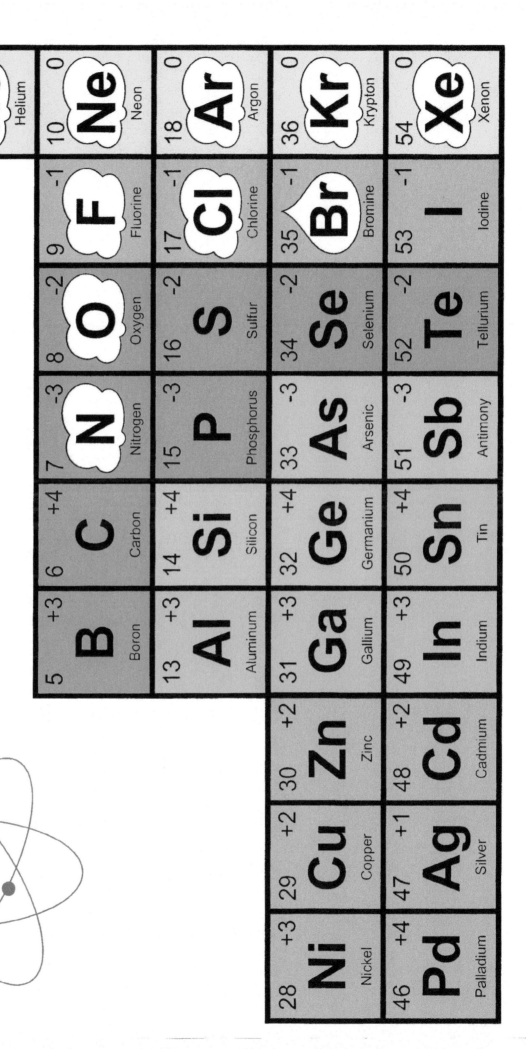

5 +3 **B** Boron	6 +4 **C** Carbon	7 -3 **N** Nitrogen	8 -2 **O** Oxygen

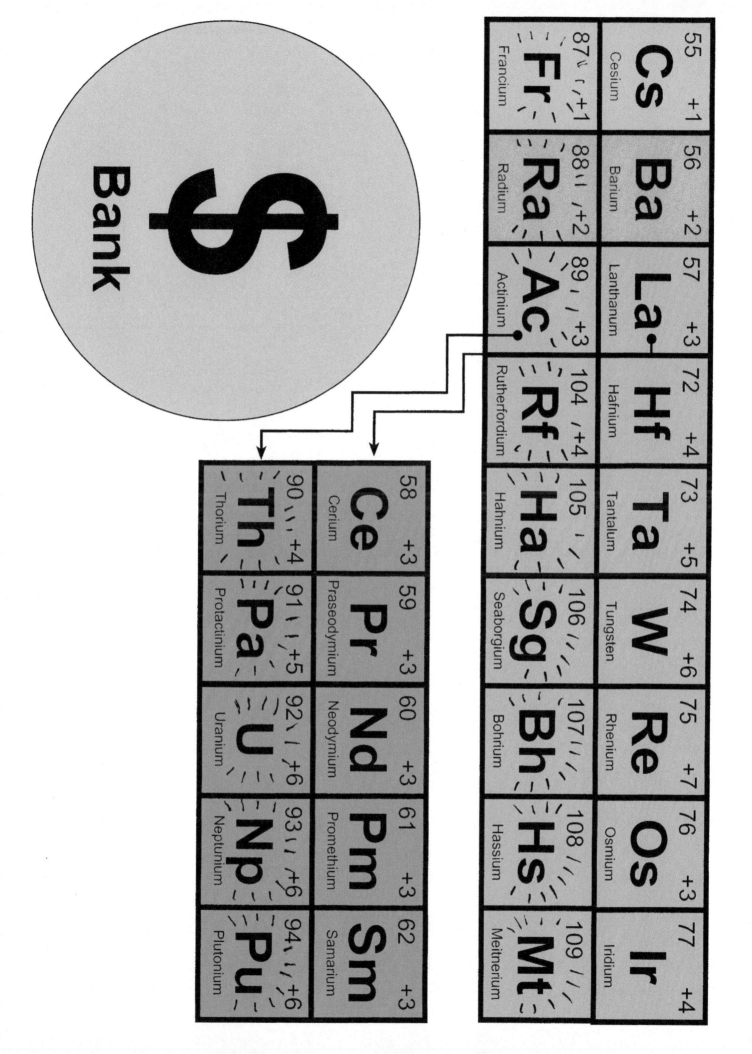

Finish

| 78 Pt +4 Platinum | 79 Au +3 Gold | 80 Hg +2 Mercury | 81 Tl +3 Thallium | 82 Pb +4 Lead | 83 Bi -3 Bismuth | 84 Po -2 Polonium | 85 At -1 Astatine | 86 Rn 0 Radon |

This top row is call the Lanthanide Series because it follows Lanthanum

Go to 72 Hf

| 63 Eu +3 Europium | 64 Gd +3 Gadolinium | 65 Tb +3 Terbium | 66 Dy +3 Dysprosium | 67 Ho +3 Holmium | 68 Er +3 Erbium | 69 Tm +3 Thulium | 70 Yb +3 Ytterbium | 71 Lu +3 Lutetium |

| 95 Am +6 Americium | 96 Cm +4 Curium | 97 Bk +4 Berkelium | 98 Cf +4 Californium | 99 Es +3 Einsteinium | 100 Fm +3 Fermium | 101 Md +3 Mendelevium | 102 No +3 Nobelium | 103 Lr +3 Lawrencium |

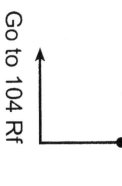

This bottom row is call the Actinide Series because it follows Actinium

Go to 104 Rf

2) CRAFT: Make a Periodic Table pillowcase

You will need: copies of the following pattern pages, clear tape, a blank pillowcase (white or a very light pastel color is best), fabric markers, glow-in-the-dark paint (if possible), and some pins to hold the pattern in place (and an iron if the instructions on your fabric markers say to use one)

What to do:
1) Copy the pattern pages onto regular paper (no need for card stock). Tape the four pages together so that they form a blank Periodic Table. Put this inside the pillowcase. You should be able to see the black lines right through the fabric. Adjust the pattern so that it is placed in the middle of the pillowcase, and pin it in place.

2) Use the fabric markers to trace over the squares. Color code the families. You don't have to use the color code shown here. You can decide what color to make each family. (If you want to add more elements, after 109, you are welcome to add them. These are what I call the "extremely silly elements" because they really don't exist. A few atoms blink in and out of existence for a millionth of a second. But you are welcome to add them to that bottom row if you want to.)

3) Write in the symbol for each element and its atomic number.

4) FUN EXTRA FEATURE: You could put glow-in-the-dark paint on the radioactive elements. GITD paint is easily obtained from any craft store and is not expensive. Look at your Periodic Table game (or find a Periodic Table on the Internet) to see which elements are radioactive. (Don't forget Technetium!)

5) Follow any ironing or washing instructions that come with your fabric markers.

TIP: One mistake that is far too easy to make is to mis-number the third row in the transition metals, forgetting to jump from Lanthanum (57) down to the lanthanide row below the main table. You might want to write the numbers 57 and 72 on the correct spaces on the paper pattern using black ink so that the numbers will show through along with all the rectangle lines. The students will see these numbers as they work and will be reminded not to forget to go down to the lanthanide and actinide rows. Another option would be to take a few minutes to look at the pattern, before they even pick up a marker. Have them notice the placement of hydrogen. (That's another common mistake—writing H on the space that is supposed to be Li.) Also, have them look at where the jumps are to the lanthanide and actinide rows, as well as noticing that the numbers stop at 109 but more can be added. (The elements past 109 are what I call the "very silly elements" because they don't really exist. An extra proton sticks to a nucleus for a split second and they declare they've discovered a new element. It's a bit ridiculous in my opinion, but some kids are fascinated by these transient elements.)

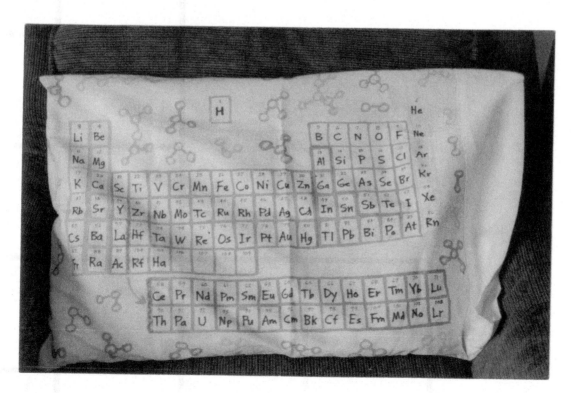

Upper Left

Hydrogen

3) GROUP GAME: "Quick Six" -- Round 2

You will need to make more cards using these additional pattern pages. Rules are the same as before, only the list of clues has expanded. Make sure the students have a Periodic Table to look at while playing, since some of the clues are about the location of the element on the Table.

Note: Some of these clues require the students to look at the atomic mass (weight) of the element. The atomic mass is listed in smaller print right under the atomic number. It is basically the number of protons and neutrons added together. Electrons are so small they add almost nothing to the total mass. The students may notice that some of the atomic masses are decimal numbers, instead of whole numbers, and they may wonder if this means that there can be fractional pieces of protons and neutrons. The reason for these decimal numbers is that scientists measured many atoms, then took a mathematical average. Since a small percentage of atoms more or less than the average number of neutrons, the average comes out to a decimal number. For example, if you weigh ten atoms of neon and get these results: 20, 20, 20, 20, 20, 20, 20, 20, 21, 21, then take the average, you will get 20.2. This is the atomic mass listed for neon. Most neon atoms have 10 protons and 10 neutrons, but once in a while you will meet a neon atom with 10 protons and 11 neutrons. (Remember, as long as it has 10 protons, it's still neon!)

When you get to the end of the clues, just start at the beginning of the list again. Add some of your own clues, too!

Atomic number has a 3 in it
Name has two syllables
Used in lasers
Has something to do with the color green
Named after someplace in Scandinavia   *(Y, Sc, Ho, Tb, Er, Yb, Hf, Th)*
Has something to do with teeth
Named after a Greek god or goddess
Is a transition metal
Starts with the letter C
Is in the same row as gold on the Periodic Table
Used in some kind of engine
Atomic number has a 5 in it
Used to make tools of some kind
Is named after a city (not a country)
Is an alkali earth metal
Is radioactive
Name has three syllables
Is used to make jewelry
Used for something that burns
Is a non-metal
Atomic mass is less than 30
Named after something in the solar system
Atomic number has a 7 in it
Is on the edge of the Periodic Table

Atomic mass is between 50 and 70
Named after Ytterby, Sweden
Is a true metal (or a semi-metal, if you have those labeled)
Is named after a country (not a city)
Used in fireworks
Has a valence of -1
Atomic number has three digits
Found in the sands of Florida and California
Is in the actinide series
Has something to do with bones
Name starts with a vowel
Is in the same row as molybdenum on the Periodic Table
Gemstones are made from it
Named after a famous scientist
Has an atomic number greater than that of tungsten
Used to color glass
Name has four syllables
Atomic number has a 0 in it
Used in steel production
Used to repair the human body in some way
Is in the same column as helium on the Periodic Table
Used in light bulbs
Has a valence of +1
Atomic mass is greater than 100
Is found as a gas in the air around us
Has something to do with eyes
Atomic number has a 9 in it
Is in the lanthanide series
Conducts electricity
Last three letters of the name are I U M
Is in the same row as iron on the Periodic Table
Has no commercial or scientific use
Is made in nuclear reactors
Name is from a Latin word
First letter of name does not match first letter of the symbol
The symbol consists of only one letter
Atomic mass is over 200
Has a valence of 0   (the noble gases)
Has an atomic number smaller than argon's number
Is completely surrounded by other elements on the table (not on an edge)

Cs 55
Cesium   132.9

Latin: "caesius" (sky blue)

- Will melt in your hand.
- Used in atomic clocks.
- Used as a "scavenger" (collector) of unwanted atoms of gas in vacuum tubes.

Ba 56
Barium   137.3

Greek: "barys" (heavy)

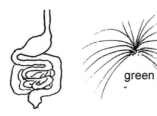

green

- Used for X-rays of digestive systems.
- Used in fireworks (green color), magnetic recording tapes, and spark plugs.

La 57
Lanthanum   138.9

Greek: "lanthanein" (to lie hidden)

lenses

- Used in telescope and camera lenses.
- Used for electrodes in high intensity lights (example: search lights).

Ce 58
Cerium   140.1

named after the asteroid Ceres

CERES

- Found in sand along the coasts of California, Florida, India and Brazil ("monazite sand").
- Used in self-cleaning ovens.
- Used in electrodes in lights.

Pr 59
Praseodymium   140.9

Greek: "prasios-didymos" (green twin)

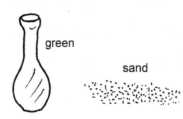

green

sand

- Found in sand along the coasts of California, Florida, India and Brazil ("monazite sand").
- Used to color glass green.
- Used in electrodes in lights.

Nd 60
Neodymium   144.2

Greek: "neos-didymos" (new twin)

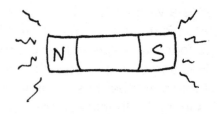

N   S

- Found in sand along the coasts of California, Florida, India and Brazil ("monazite sand").
- Used to make very strong magnets.
- Used to color glass and to make rubies.

Pm 61
Promethium   147.0

named after Greek god Prometheus

- Is a synthetic element made in nuclear reactors.
- Can be a source of X-rays in portable X-ray machines.

Sm 62
Samarium   150.3

named after the mineral "samarskite" which was named for Col. Samarski, a Russian army engineer

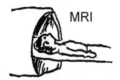

MRI

sand

- Found in sand along the coasts of California, Florida, India and Brazil ("monazite sand").
- Used in magnets for MRI machines, and in infra-red absorbing glass.

Eu 63
Europium   151.9

named after Europe

- Used to make red color in televisions.
- Used in mercury lamps and energy-saving fluorescent bulbs.
- Used to identify counterfeit Euros.
- Used to study formation of igneous rocks.

Gd 64

Gadolinium   **157.2**

named for chemist Johann Gadolin

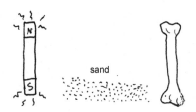

- Found in sand along the coasts of California, Florida, India and Brazil ("monazite sand").
- Used in magnets and TV tubes.
- Used to diagnose osteoporosis.

Tb 65

Terbium   **158.9**

named after Swedish village of Ytterby

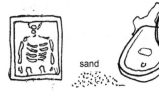

- Found in sand along the coasts of California, Florida, India and Brazil ("monazite sand").
- Used in TV tubes and X-ray screens.
- Used in metal alloys for CD players.

Dy 66

Dysprosium   **162.5**

Greek: "dysprositos" (difficult to obtain)

- Found in sand along the coasts of California, Florida, India and Brazil ("monazite sand")..
- Used in TV tubes, mercury lamps, and magnets inside CD players.

Ho 67

Holmium   **164.9**

named for Stockholm, Sweden

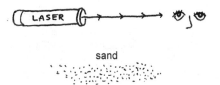

- Found in sand along the coasts of California, Florida, India and Brazil ("monazite sand").
- Used in eye-safe medical lasers.
- Used to color glass.

Er 68

Erbium   **167.3**

named after Sweidish village of Ytterby

pink

- Used in alloys with vanadium, to make the texture less brittle (easier to shape).
- Used for pink coloring in glass.
- Used to make artificial gemstones.
- Superconducts at low temperatures.

Tm 69

Thulium   **168.9**

Thule is the ancient name for Scandinavia

- Found in sand along the coasts of California, Florida, India and Brazil ("monazite sand").
- Used in lasers and in medical imaging.
- Is very rare.

Yb 70

Ytterbium   **173.0**

named after Swedish village of Ytterby

- Found in sand along the coasts of California, Florida, India and Brazil ("monazite sand").
- Used in dentures (artificial teeth).
- Is added to stainless steel to improve strength.

Lu 71

Lutetium   **174.9**

Lutetia is the ancient name for Paris

- Found in sand along the coasts of California, Florida, India and Brazil ("monazite sand").
- Is the only naturally-occurring element discovered in America.
- Used in temperature-sensing optics.

Hf 72

Hafnium   **178.5**

Hafnia is the ancient name for Copenhagen

- Usually found with zirconium.
- Used in nuclear submarines and nuclear reactors.
- Used as a gas "scavenger" (collector) in vacuum tubes (to get rid of unwanted atoms of gas).

Ta 73	W 74	Re 75

Ta 73
Tantalum 180.9
named after the Greek god Tantalus

weights

capacitors

- Used to repair bones, especially in the skull.
- Used to make tools and weights.
- Used for capacitors in electronics.

W 74
Tungsten 183.8
Swedish: "Tung stem" (heavy stone)
Used to be called Wolframite

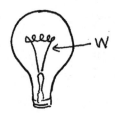 —W

- Used for filaments in light bulbs.
- Used for high-speed cutting tools.
- Has the highest melting point of all the metals.

Re 75
Rhenium 186.2
Latin: "Rhenus" (Rhine River)

- Used in alloys, especially for electrical switches and contacts.
- Used for high-temp thermometers.
- Used for oven filaments.

Os 76
Osmium 190.2
Greek: "osme" (smell)

- Used in pen points and compass needles.
- Mixed with platinum and iridium to make alloys.
- Is the most dense element, twice as dense as lead.

Ir 77
Iridium 192.2
Latin: "iris" (rainbow)

  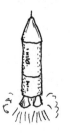

- Iridium salts are highly colored.
- Used in helicopter spark plugs, hypodermic needles and rocket engines.
- Is often mixed with platinum.

Pt 78
Platinum 195.1
Spanish: "platina" (silver)

- Used in jewelry and dentistry.
- Used in the petroleum and electronic industries.
- Most platinum comes from South Africa and Russia.

Au 79
Gold 196.9
Old English: "gold"
"Au" comes from Latin: "aurum"

- Used for coins, jewelry, dentistry, and electrical parts that need to conduct electricity.
- Used as a reflective coating on the outside of large glass windows.

Hg 80
Mercury 200.6
named after the Roman god Mercury

- The symbol **Hg** comes from the Latin "hydragyrum" meaning "liquid silver."
- Used in thermometers, barometers, and street lights.
- Found primarily in the mineral ore "cinnabar," mined in Spain and Italy.

Tl 81
Thallium 204.4
Greek: "thallos" (green twig)

- Looks like lead and is poisonous.
- Was once used in insecticides.
- Used to diagnose heart disease.
- Used in infrared detectors.

Pb 82 **Lead** 207.2 *Ancient Anglo-Saxon: "lead"* *"Pb" comes from Latin: "Plumbum"* 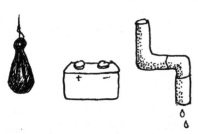 • Used for fishing weights, in batteries, and for protection against X-rays. • Romans used lead for their water pipes.	**Bi** 83 **Bismuth** 208.9 *German" "weisse masse" (white mass)*  • Used in stomach medicines such as Pepto-bismol® • Used in indoor sprinkler systems (fire safety for commercial buidings). • Used in the manufacturing of rubber, fuses, and cosmetics	**Po** 84 **Polonium** 210 *named after Poland*  • Discovered by Marie Curie, who was born in Poland. • Is very radioactive. Can be used as a source of radiation.
At 85 **Astatine** 210 *Greek: "astatos" (instable)* 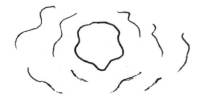 • Very little is known about this element. • The total amount of astatine that exists is estimated to be only about an ounce! • Is radioactive.	**Rn** 86 **Radon** 222 *named after the element radium*  • Is the heaviest gaseous element. • It is radioactive and probably causes lung cancer. • Used in earthquake prediction.	**Fr** 87 **Francium** 223 *named after France*  • Discovered in France. • Is very active. • Comes from the decay of uranium and thorium. • Is too unstable to be used for anything.
Ra 88 **Radium** 226.0 *Latin: "radius" (ray)*  • Discovered with the spectrometer, as an impurity in uranium ores. • Was once used to make glow-in-the-dark watches. • Can be used to make radon, for use in medical procedures.	**Ac** 89 **Actinium** 227 *Greek: "actinos" (ray or beam)*  • Is radioactive. • Comes from the decay of uranium and thorium. • No commercial use.	**Th** 90 **Thorium** 232 *after the ancient Scandinavian god Thor, god of lightning and thunder*  • More common than uranium. • Used as a source of electrons in some electronic devices. • Used in the "mantles" of camping lanterns (that little bag-like thing that glows)

Pa 91
Protactinium 231

Greek: "protos" (first), plus "actinium"

- Was given this name because it always decays into actinium.
- Not much is known about it.
- Has no commercial use.

U 92
Uranium 238

named after the planet Uranus

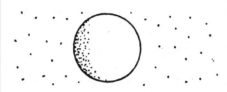

- Is radioactive.
- Was discovered just after Uranus was.
- Used as fuel in nuclear reactors.
- Depleted uranium (which is much less radioactive) is used to color glass and to make metals for military vehicles.

Np 93
Neptunium 237

named after the planet Neptune

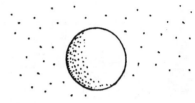

- Is radioactive.
- Is produced as a by-product of nuclear fission.
- Very small quantities of naturally-occurring neptunium have recently been discovered in uranium ores.

Pu 94
Plutonium 242

named after Pluto

- Is made from uranium inside "breeder" nuclear reactors.
- Used in nuclear weapons.
- Was used to power the lunar modules.
- The element barium was almost named plutonium!

Am 95
Americium 243

named after America

- Is radioactive.
- Used in smoke detectors.
- Used in crystal research.
- Used as a source of neutrons.

Cm 96
Curium 247

named after Marie Curie

- Is radioactive.
- Used in pacemakers in heart, and also in ocean buoys.
- Has been used as an energy source on space missions.

Bk 97
Berkelium 247

named after Berkeley, California

- Is radioactive; was made in Berkeley, Ca.
- Has no commerical use.
- BkCl3 (berkeliium trichloride) was the first compound to be made with this element. The quantity produced was very small-- only.000000003 of a gram!

Cf 98
Californium 251

named after California

- Is radioactive.
- Can be used as a portable source of neutrons.
- Named after California because that's where it was made/discovered.

Es 99
Einsteinium 252

named after Albert Einstein

- Discovered during the investigation of debris from the first atomic bomb.
- Extremely radioactive and unstable.
- Einstein is famous for his equation that shows the relationship of matter to energy ($e=mc^2$).

Fm 100
Fermium   257

named after Enrico Fermi

- Discovered during investigation of the debris from the first atomic bomb.
- Extremely radioactive and unstable.
- No commerical use.
- Fermi was a physicist who studied atomic structure and radioactivity.

Md 101
Mendelevium   256

named after Dmitri Mendeleyev

- Radioactive and very unstable.
- Made in nuclear reactors.
- No commerical use.
- Mendeleyev invented the Periodic Table.

No 1[0]
Nobelium

named after Alfred Nobel

- Very radioactive and very unstable.
- Made in nuclear reactors.
- No commerical use.
- Alfred Nobel established the Nobel Prizes.

Lr 103
Lawrencium   262

named after Ernest Lawrence

- Is radioactive and very unstable. It only exits for a few minutes after it is created.
- Lawrence was the inventor of the cyclotron machine that was used to discover elements heavier than uranium.

Rf 104
Rutherfordium   261

named after Ernest Rutherford

- Is very radioactive and unstable.
- Is made in nuclear reactors.
- No commerical use.
- Rutherford was a famous physicist.

Db 105
Dubnium   262

named after Dubna, Russia

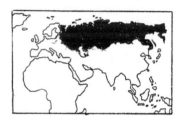

- Was made in a reactor in Russia.
- Is very radioactive and very unstable.
- Only exits for a few minutes.

Sb 106
Seaborgium   263

named after Glenn T. Seaborg

- Is very radioactive and unstable
- Is made in nuclear reactors.
- No commerical use.
- Only exits for a few seconds.
- Seaborg and his team discovered Pu, Am, Cm, Bk, Cf, Es, Fm, Md and No.

Bh 107
Bohrium   262

named after Niels Bohr

- Extremely radioactive and unstable.
- No commercial use.
- Only exists for a fraction of a second.
- Niels Bohr figured out atomic structure and also studied the nature of light.

Hs 108
Hassium   265

named after Hesse, Germany

- Extremely radioactive and unstable.
- Made in nuclear reactors.
- No commericial use.
- Only exists for a fraction of a second.

ACTIVITY IDEAS FOR CHAPTER 5

the Seaside" (about the discovery of iodine and bromine)
e Cow Who Wouldn't Drink" (about the discovery of magnesium)

don't want to actually perform these skits, they can be done as "Reader's Theater" with no props or rehearsals. The participants simply sit in a circle and read their lines dramatically. A sound-effect person can be added at the director's discretion.

You will need: copies of the scripts (following), one for each actor.

"By the Seaside"
        If you want a few props for "By the Seaside," here are some suggestions: a table and chair, artificial seaweed (green paper is fine), beach hat and sun glasses, some pots, a bottle marked "sulfuric acid," a small clear container with red liquid in it, a newspaper, a Periodic Table. (You can also do the whole skit as pantomime, without props. If the actors pantomime well, the audience's imagination will fill in the details.)

If you want to use signs to show the audience:
    "Western coast of France, 1811"
    "Several hours later..."
    "Several months later..."
    "Heidelberg, Germany, 1825"
    "ACT I"
    "ACT II"
    "The end"

"The Cow Who Wouldn't Drink"
        If you would like to use props for "The Cow Who Wouldn't Drink," you might want to gather:
a mug or cup, a bell for the cow, hat and shovel for John, an opaque container, a flame made of construction paper, and a pretend electrolysis machine made like this:

If you would like to use signs to show the audience:
    "Epsom, England, 1618"
    "Several months later..."
    "A year later..."
    "Humphry Davy's lab, England, 1855"
    "ACT I"
    "ACT II"
    "The end"

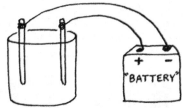

The container can be any clear container, the electrodes can be made of anything (even pencils will do), the wires can be made of wire or string or yarn, and the battery can be a large lantern-type battery, a box labeled "battery," or (if you want to be historically accurate) a facsimile of a Voltaic pile or Leyden jar (search Internet for images). Don't be overly worried about accuracy. Kids' imaginations will fill in the gaps.

NOTE: Humphry Davy was not only a researcher, but also a famous science lecturer of the 1800s. People would go to his lectures like we go to movies. (He was sort of the "Bill Nye the Science Guy" of his day, but for adults. He was interesting and funny, not just scientific.) In this skit he has a long monologue because he is giving one of his famous public lectures.

"By the Seaside"

A skit about the discovery of iodine and bromine

NOTE: This skit does not quote what these individuals actually said, as we have no record of their exact words. The conversations are fictional, but the general facts presented are true.

Characters:
- Bernard Courtois ("Cur-TWAH"), a French chemist
- Professor Gmelim, chemistry professor at the University of Heidelberg
- Carl Lowig, a freshman chemistry student
- Narrator (or sign-holder, if you are performing the skit using signs)

* * * * * * * * * * * * * *

SCENE 1

Narrator: Bernard Courtois, a French chemist, is beachcombing on the Western coast of France in 1811.

Courtois: What a lovely day at the beach! I'm so glad my work brings me out here occasionally. *(As he walks along he collects seaweed.)* Being a chemist, you'd think I do nothing but work in my laboratory. But I do a lot of work with two of the most recently discovered elements: sodium and potassium. To get pure sodium and potassium, I burn seaweed, then treat it with acid to get rid of the stuff that I don't want in it, leaving me with the sodium and potassium I need. Well, I think I have about enough seaweed for now. I think I'll go back to the lab and get to work.

Narrator: Several hours later...

Courtois: This is the point at which I must pour in some acid to get rid of all the impurities. I'll just open the bottle and carefully pour in a tiny amount. Aahh. What's that on my shoulder? Mr. Spider, don't scare me like that. Get off! Get off! *(If you are performing, jerk around trying to get the spider off. While you are doing so, make sure you accidentally pour too much acid in the pot.)* Oh no! Look what I did! That was almost the whole bottle! Who knows what could happen?! Eeww! Eeww! My concoction is turning purple! It smells awful, like chlorine, only worse! Oh, no! Purple smoke! Ack! Gasp! I'm getting out of here!

Narrator: Several hours later Courtois came back in the room and discovered something curious all over the table.

Courtois: What is this stuff? It's all over. It must have condensed from that purple smoke. I've never seen anything like this before. I must try to find out what it is.

Narrator: Several months later, Courtois makes the announcement of his new discovery.

Courtois: *(loudly, like you are announcing!)* And so, I would like to announce to the world the discovery of a new element: Iodine! I've chosen this name because the Greek word for purple is "iodes." *("i-o-deez")* I have started a small factory to produce iodine. I am sure it will turn out to be good for something someday, and we'll need a lot of it. So, if you need a job and don't mind wearing a purple uniform, come to my factory! Thank you.

SCENE 2

Narrator: We are in a classroom at a university in Heidelberg, Germany, in 1825.

Professor Gmelim: And that concludes today's chemistry lecture. I hope you have enjoyed your first day here at Heidelberg University. I'll see you here tomorrow at 9:00 AM. Class dismissed.

Carl Lowig: Professor! Professor! Can I show you something really neat? It will only take a second. Look at this vial of red liquid. Take a look at it and tell me what you think it is.

Professor: *(after examining it)* Hmm... I don't know. Maybe a sniff of it will give me a clue. (Sniff) Agh! This stuff is strong! It reminds me chlorine. Yuck!

Carl: Do you think it might be chemically related to chlorine?

Prof: I can't rightly say what it is. Where did you get this stuff?

Carl: I cooked it up in my laboratory at home. I put in a whole bunch of stuff. We had just come back from our vacation at the seaside so I threw in all kinds of rocks and seaweed, and even some sea water. Then I added acids from my chemistry set.

Prof: Whatever it is, it sure smells strong! Could you make some more of this stuff during the semester? Maybe over Christmas vacation we could find time to analyze it and find out what it is. I don't think I've seen anything like this before. I wonder, though, since it smells so much like chlorine, if it might be the missing element on the Periodic Table, right under chlorine and above iodine. The French guy who discovered iodine reported that it smelled like chlorine, too. It's worth checking out, anyway.

Narrator: Several months later..

Carl: Professor, I've managed to make a bunch more of that smelly red stuff, but I'm afraid it doesn't matter now.

Prof: Why not?

Carl: I just read in the newspaper that a new element has been discovered. The description of it fits my red stuff exactly. *(hands paper to Prof.)*

Prof: Hmm... I'm afraid you might be right, Carl. This sounds exactly like your mysterious red liquid. Here's the clincher: "The new element has been named "bromine" from the Greek word "bromos" meaning "stench." I guess they thought it smelled bad, too, eh? Well, you win some and you lose some. Good luck with the rest of your semester, Carl.

Carl: Thanks, Professor.

Narrator: Don't feel too sorry for Carl. He's going to have a brilliant career as a chemist. Any guy who can cook up a new element over summer vacation is bound to be a success!

"The Cow Who Wouldn't Drink"
A skit about the discovery of magnesium

Characters:
- narrator
- John Epsom
- a cow
- a business man
- Humphry Davy, a famous scientist (of the 1800s) who gave public lectures

SCENE 1

Narrator: We are in Epsom, England, in the year 1618.

John Epsom: My name is John Epsom. My father's name was John Epsom. My grandfather's name was John Epsom. My great-grandfather's name was... James Epsom. Our little town is called Epsom because my great-great-great-great-great-great-great-grandfather, Thomas Epsom, founded this town. Today is an exciting day on the Epsom family farm. I've just finished digging the new well. It's much closer to the barn than the old well, so I won't have to walk so far to water the cows every day. It took me 30 days to dig this well, but it will be worth it. Well, ol' Bessie, ready to take the first drink? Step right up... there you go...

Narrator: The cow walked to the edge of the well, put her head down, sipped the water and..

Cow: Mooo! Moooo! Mooo! *(violent moos, as if the cow is saying "Yuck!")*

John: What?!! You won't drink? It took me 30 days to dig this well. Drink! Drink!

Cow: Mooo! *(violent mooing again, cow shakes its head "No")*

John: What a crazy cow! Doesn't know good water even when you lead her right to it! I think I'll take a sip myself, then. This water is the... *(takes a sip)*.. MOST HORRIBLE STUFF I'VE EVER TASTED! AGH!

Narrator: Several months later...

John: Hello, again. Since you were last here, I've made a most amazing discovery. Remember that well I dug for my cows? How could you forget, right? It was filled with the most horrible-tasting water in the world. Well, I decided to do some more excavating to see if I could find out why it was so bad. I spent a couple of weeks mucking about, digging around in it. Since I am right-handed, my right arm was the one I had down the well most of the time. I noticed that the skin on my right hand started looking a lot better than the skin on my left hand. Any scratch or rash I had on my right hand healed twice as fast as the ones on my left hand. So I started taking a bottle of well water home for my wife to try out. She said she found the same thing. I think there might be money in this discovery someday...

Narrator: A year later...

Businessman: You've gotta have a catchy motto, John. How about this one: "Epsom Salts, your key to good health." Or how about, "Epsom's famous healing water. If you don't drink it, don't blame us for your ills."

John: You make a better businessman than advertiser. How are sales going?

Businessman: The good news is that sales are better than ever. The bad news is that we'll never be able to keep up with the amount of water we have to transport. I suggest that we bottle those little crystals we find around the well, instead of bottling the water. We could save ourselves huge shipping costs.

John: Great idea! I think those little crystals are the key to what makes the water so healthful. We could easily ship those crystals all over the world. Well, what are we waiting for?

SCENE 2

Narrator: We now go to the laboratory of Humphry Davy, in England in the year 1855.

Davy: Hello, everyone. I am Professor Davy. Today's chemistry demonstration is about electrolysis. I have here a solution of magnesium oxide and mercuric oxide. I will now put the two electrodes into the solution. One electrode is positive and the other is negative. What should happen is that the negative electrode will attract the magnesium and mercury atoms. There! A clump of solid magnesium and mercury atoms! Now I just heat this clump and the mercury atoms will turn into vapor and fly off into the air... leaving... pure magnesium!

Ladies and gentlemen, if you are a student of English history, you will remember a story about a farmer who discovered the water in his well to be bitter. He never knew what was in the water, even though he made a fortune on it. I have solved his mystery. I believe his water had magnesium in it. The magnesium gave his water its excellent healing properties as well as its bitter taste. Magnesium seems to be an element that our body needs, it just doesn't taste very good. Today we call his little crystals "Epsom salts."

When I first did this experiment, I had no idea what these atoms were. I was the first person in history to produce the this pure element, magnesium. While I was trying to think of a good name for this new element, my lab assistants started calling it "magnesium." I said the name "magnesium" would be far too confusing for people, since we already have an element called "manganese." Chemistry students in the future will find it difficult to keep "magnesium" and "manganese" straight. But no! No one listened to me! They went right on calling it magnesium. So, now that everyone calls it that, we are stuck with it. Well, at least I got to name sodium, chlorine, and potassium.

Thank you for coming to my lecture today, ladies and gentlemen.

2) LAB DEMO: Make a (harmless) smoke bomb out of potassium nitrate (KNO₃)

Potassium nitrate, KNO_3, is one of the ingredients in gun powder. Here, you can use it to make a harmless smoke bomb.

You will need: sugar, potassium nitrate (sold as tree stump remover in garden supply centers), pan and spoon, cotton string, candle, aluminum foil and/or thin cardboard tubes (for larger bombs)

What to do:
1) Before you start, make some wicks (fuses). Cut short pieces of cotton string and let candle wax drip on them. When the wax cools slightly, rub wax into string. (If you can't make wicks, that's okay, they will still light without them, you just have to be more careful when lighting them.)
2) The recipe for the ingredients is 3 parts KNO_3 to 2 parts sugar. Stir them together, then put into metal pan.
3) Heat these ingredients on LOW heat, slowly, just warm enough so that they melt together. DO NOT USE a heat source with OPEN FLAME. Use an electric heat source. (Some people also add a bit of paraffin. You can compare recipes online by typing key words into Google.)
4) Stir until it looks like lump of soft caramel candy.
5) You can put small lumps of it onto aluminum foil to cool (as if they were cookies) or you can put a lot of it into a thin cardboard tube such as a toilet paper tube.
6) Put wicks (fuses) in. (However, most instructions say you can also light them directly. Fuses just give you more time to back away before the burning starts. And they do burn energetically!)
7) Use smoke bombs outside, never inside. Follow common sense safety precautions when lighting. Avoid breathing too much of the smoke.

HINT: Watch the demo video on the YouTube playlist, showing what they do when they ignite.

3) LAB DEMO: An edible chemical reaction

This lab uses a compound that contains sodium (Na), an alkali metal.

You will need: sugar, baking soda, and citric acid (available in the canning supplies section)

Here is the recipe:
> 2 teaspoons citric acid
> 1 teaspoon baking soda ($NaHCO_3$)
> 6 teaspoons sugar

What to do: Grind these ingredients together and mix thoroughly. Put some of this mixture on your tongue. What do you feel?

What is happening: The chemical reaction in this experiment produces bubbles of carbon dioxide. That is the fizzing feeling in your mouth—carbon dioxide bubbles being formed. Carbon dioxide bubbles are the "fizz" in soda pop. The citric acid is an acid (obviously). The baking soda is the opposite of an acid: a base. (Another word for a base in an "alkali." That should sound familiar.) Two things are being produced in this reaction: water and carbon dioxide.

How is water produced? Acids, by definition, are protons donors. In other words, they easily lose hydrogen ions, H+. A hydrogen ion is a proton with no electron. A base is a substance that either donates OH- ions itself or causes OH- ions to become available. When you put an H+ together with an OH-, you get a water molecule, H_2O. Therefore, acids and bases neutralize each other and form harmless water molecules.

How is carbon dioxide produced? If you look at the formula showing how the atoms are rearranging, you will see that after the Na and H are broken off, an oxygen leaves the CO_3 to make CO_2.

$$C_6H_8O_7 + NaHCO_3 \rightarrow Na_3C_6H_5O_{7} + H_2O + CO_2$$
Citric acid, $C_6H_8O_7$ combines with $NaHCO_3$ to form sodium citrate, $Na_3C_6H_5O_7$ plus H_2O and CO_2.

4) LAB EXPERIMENT using magnesium: Make some Mg(OH)$_2$

"The nurse brought Mg(OH)$_2$ and MgSO$_4$" (from the Chemical Compounds Song in chapter 1)
In this activity you can make Mg(OH)$_2$ [milk of magnesia] by using MgSO$_4$ [epsom salts].

NOTE: This is not a super exciting experiment. The final result is just that the liquid gets cloudy. If you have students who
       need big "wow" experiments to maintain their interest, you might want to skip this one.

You will need: epsom salt (MgSO$_4$), liquid cleaning ammonia (which is actually ammonium solution, NH$_4$(OH), water (H$_2$O), a
       small clear container and a spoon
NOTE: Bottles that are labeled "ammonia" are actually ammonium solution, NH$_4$(OH). Ammonia, NH$_3$, is a gas. When you
smell the ammonium solution, molecules of ammonia, NH$_3$, come out of the solution and go up your nose (and sting!).

What to do:
   1) Fill the container half full with water. Stir in 2 spoons of epsom salt and stir the epsom salt until dissolved.
(HINT: Using hot water will help the epsom salt crystals to dissolve faster.)
   2) Add 2 spoons of ammonium solution, but don't stir.
   3) Observe the solution for 5 minutes. Does the liquid change?
   4) Let it sit for several hours or overnight and observe again. Has anything changed?

What is happening:
       The atoms are rearranging themselves. They are deciding to leave the molecules they used to be a part of and form
new ones. This rearranging of molecules causes changes that you can see. You should see a cloudy mat appearing near the
surface (which is actually the "milk" of magnesia) and possibly white dots falling to the bottom. The white dots are called
precipitates.
       Here is the formula for the chemical reaction, but it is not important that the students understand it completely. It's
enough to know that Mg would rather be attached to the OH's, so it drops the SO$_4$ to do so. The SO$_4$ is fine with attaching to 2
NH$_4$'s. (Remember Mg(OH)$_2$ from the Chemical Compounds Song?)

$$MgSO_4 + 2\ NH_4(OH) \rightarrow Mg(OH)_2 + (NH_4)_2SO_4$$

5) CRAFT: Make a model of an ionic compound

You will need: some table salt, a magnifier, some marshmallows (or a substitute such as grapes or clay balls, depending upon
       what you want to do with the molecule when it is finished) and some toothpicks.

What to do: Ionic compounds often form a shape called a lattice. A lattice is a regular arrangement of atoms into a certain
shape. In NaCl, the shape is cubic. Use the magnifier to look at some salt crystals. Are they basically cubic? They may not be
perfect cubes, but they should look cubic rather than round or triangular or hexagonal. The salt crystal is cubic because the
atoms of sodium and chlorine are lined up in squares. Use this picture as a guide to create your own salt crystal lattice model.

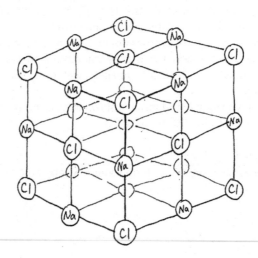

6) ART PROJECT: Element Trading Cards

Your students might collect baseball or football cards, but now they can collect chemical element cards, too! I recommend making this at least a three-session project, doing a few cards each week. During this first work session, choose members of the alkali metals, the alkali earth metals, and the halogens. You can use the "Quick Six" cards to help you find atomic weights and facts about the elements, or you might want to use Internet resources, or books from the library—it's up to you. (If you want an all-in-one site for information, try the key words "interactive Periodic Table.")

You will need: copies of the trading card pattern sheet printed onto white card stock. If you want to have the blanks on the back for adding information (instead of asking students to do reports) make the copies double-sided.   Make enough copies so that students will be able to make at least a dozen cards if they want to. Each student will need one copy of the card holder, also printed onto card stock, although you can use colored stock if you want to.

Your students will need: scissors, white glue and colored pencils, plus anything extra you want to provide such as glitter, metaillic markers, gel pens, etc.

The "team" the element is on is the family group to which it belongs (alkali metals, non-metals, etc.) The number is the atomic number. Color in the correct space for its position on the field. The artwork can be totally creative. The students can draw something that has the element in it, or they can make up a cartoon player. Below are some cards that students have made in the past.

Sample cards showing how to use bright colors.

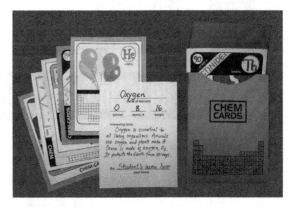

The card holder (pattern on next page) provides a great place to store your finished Chem Cards.

(TIP: Berol Prismacolor colored pencils are really good for this. You can get them on Amazon, but don't buy the "student pencils" because they are not as good. Prismacolors are not cheap, but they are worth every penny.)

137

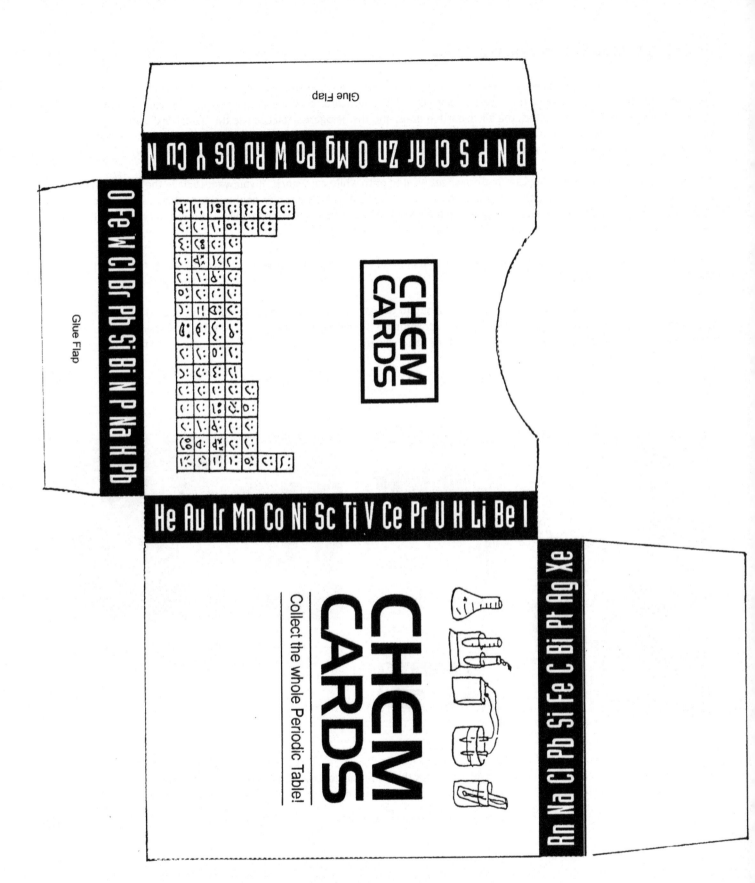

Holder for trading cards. Copy one per student onto heavy card stock (any color).

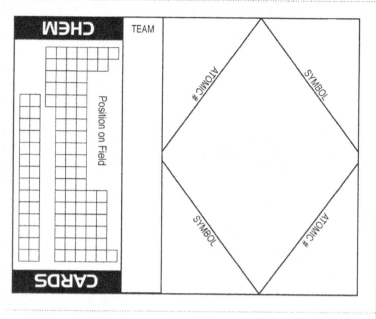

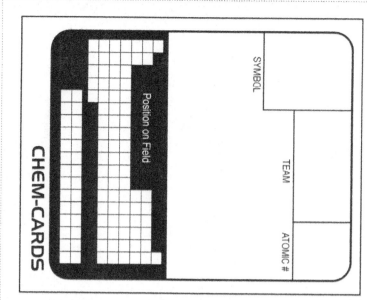

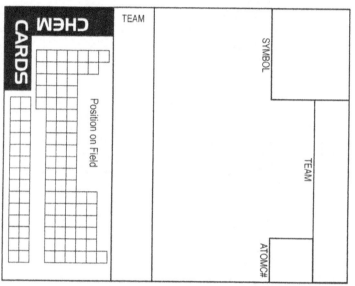

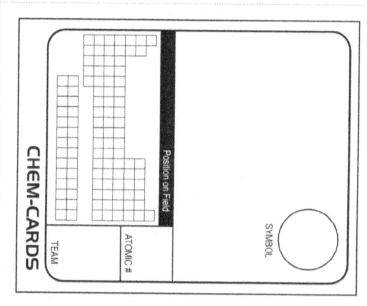

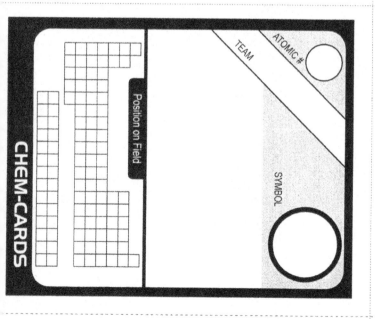

Pattern for trading cards

Copy onto white card stock.

139

____
name of element

______ symbol ______ atomic # ______ weight

Interesting facts:

Card designed by: ______________________
your name

____
name of element

______ symbol ______ atomic # ______ weight

Interesting facts:

Card designed by: ______________________
your name

____
name of element

______ symbol ______ atomic # ______ weight

Interesting facts:

Card designed by: ______________________
your name

____
name of element

______ symbol ______ atomic # ______ weight

Interesting facts:

Card designed by: ______________________
your name

____
name of element

______ symbol ______ atomic # ______ weight

Interesting facts:

Card designed by: ______________________
your name

____
name of element

______ symbol ______ atomic # ______ weight

Interesting facts:

Card designed by: ______________________
your name

Copy this onto the reverse side of cards.

7) REPORT on an element

If you have students who are not interested in making trading cards, yet need an incentive to do research on specific elements, you may want to copy the following pattern page and assign the student(s) to read an article or a book on an element and then answer the questions in the boxes on the report sheet.

NOTE: This is the same form that appears in the student text at the end of chapter 7.   You can wait until then, or go ahead and do one now. Or skip it both times—it's up to you.

8) VIDEOS about the elements (streamable and free)

If your student(s) are really intrigued with the elements at this point and want more information, I suggest using the "Periodic Table of Videos." Some geeky professors in Nottingham, England, made a whole series of videos about each element on the Periodic Table. The videos range from 2 minutes to 8 minutes long. You'll get more than you want to know about some of the elements, but you'll also see difficult and/or dangerous experiments that you could never do at home. The videos are highly recommended as being well worth your time. They make a fabulous supplement to this curriculum. You could even use them for your research in making the trading cards. Watch a video, then make a card for that element. (You may have watched their videos on the noble gases, posted on The Elements playlist.) They've just put up a new web site: **www.periodicvideos.com**. The home page is a Periodic Table with clickable element squares. Just click on any element to see a video about it. If you'd rather access these videos through YouTube, they are posted there, too, both as individual videos and as a channel.

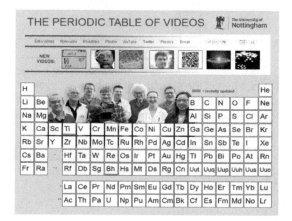

Sir Martyn Poliakoff is the host of these videos, but several other "lab guys" are regulars, too.

_____ (symbol box)

name of element

atomic mass   number of protons   number of neutrons   number of electrons

atomic number

At standard temperature and pressure (STP), this element is a:

☐ solid
☐ liquid
☐ gas

Where did this element get its name?

At what temperature will this element boil?

At what temperature will this element melt or freeze?

What group does this element belong to?

☐ alkali metals
☐ alkali earth metals
☐ transition metals
☐ true metals
☐ semi-metals (metalloids)
☐ non-metals

Is this element found in the Earth's crust? ☐ yes ☐ no
If so, where?

☐ rocks   ☐ dirt   ☐ lava
☐ water   ☐ gemstones
☐ sand   ☐ _____

When was this element first discovered?

Who discovered it?

Is this element ever found all by itself (not part of a compound)? ☐ yes ☐ no

What color is this element? (Or, if it is never found by itself, what color is its most common compound?)

Other colors?

Is this element used in industry? ☐ yes ☐ no
If so, what is it used for?

Is this element found in the human body? ☐ yes ☐ no
Is it part of the structure of the body? ☐ yes ☐ no
Can this element be harmful to the body? ☐ yes ☐ no
If it is harmful, how might you ingest it or come into contact with it?

Is this element used in any art or craft? ☐ yes ☐ no
What type of art uses it?
☐ painting   ☐ sculpture   ☐ pottery
☐ printing   ☐ coins   ☐ jewelry
☐ _____

Is this element used in medicine or dentistry? ☐ yes ☐ no
If so, how is it used?

Give one historical fact about this element other than the date of its discovery:

Draw a picture of a molecule containing this element:

What do you think is the most interesting fact about this element?

Name of molecule:

ACTIVITY IDEAS FOR CHAPTER 6

1) LAB DEMO: Something oxygen can do

You will need: food colorings and colored paper, bleach (NaClO), paper towels and napkins, water, eye dropper if possible

What to do:
  1) Put a drop or two of bleach onto each sample of colored paper. See what happens.
  2) Put a drop of red food coloring onto a wet napkin, wait half a minute until it spreads out, then put just TWO drops of bleach in the center of the food coloring area. Watch for one minute. You should see a clear area forming in the middle of the food coloring spot. It will take a few minutes for the bleach to make it totally white.

What is happening?
      The oxygen atom is coming off the NaClO and going into the dye molecule. The dye molecules have an arrangment of atoms that just happen to reflect red light. When the oxygen joins the molecule, it alters the structure of the molecule just enough that it no longer reflects red light. The red dye molecules did not "go away." They are still there! They are just not reflecting red light any more.

2) CRAFT: Make some delicious covalent molecules

You will need: some colored marshmallow and toothpicks
NOTE: You can use something other than marshmallows if you need to, such as grapes, cheese cubes or bits of dried fruit. You will need six different items to represent six types of atoms: H, O, C, S, N and Cl.

Here are the covalent molecules to make:

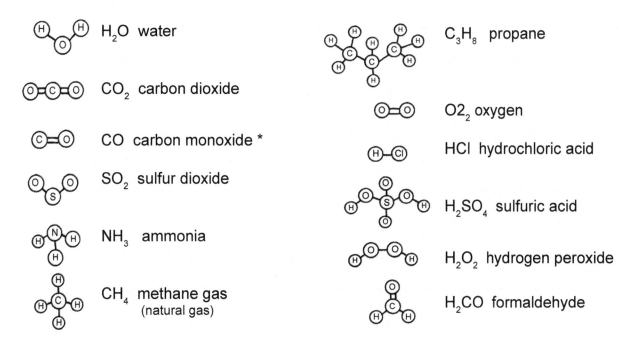

H_2O water

CO_2 carbon dioxide

CO carbon monoxide *

SO_2 sulfur dioxide

NH_3 ammonia

CH_4 methane gas (natural gas)

C_3H_8 propane

$O2_2$ oxygen

HCl hydrochloric acid

H_2SO_4 sulfuric acid

H_2O_2 hydrogen peroxide

H_2CO formaldehyde

      Here's a group game to try after you've practiced making the molecules. Write the names of the molecules on slips of paper and put them into a cup. Have all the raw materials in the middle of the table. One player draws out a molecule and lays it down for all players to see. Ready, set, go! The first player to assemble the molecule correctly wins. (The prize for winning could be eating your molecule if you are working with edibles.)

* NOTE: Sometimes carbon monoxide is shown with a triple bond (three lines).

3) LAB DEMO: Testing for oxygen (the way Priestly and Lavoisier did)

In this lab, you will perform the classic test for oxygen, the very same test that scientists like Priestly and Lavoisier used to determine if oxygen (or "de-phlogisticated air," as they called it) is present.

You will need: hydrogen peroxide (a first aid product used to clean wounds), yeast, a tall glass jar (the narrower the better) or test tube, a wooden coffee stir stick (or other long, thin piece of wood), some matches

What to do:
1) Pour some hydrogen peroxide (H_2O_2) into a tall glass jar or a glass test tube. The narrower the glass container is, the less peroxide you will need. You want to have some liquid at the bottom, but still have a long air space in which O_2 can collect.
2) Add about half a teaspoon of yeast. Stir and then let it sit for a few minutes. Watch for bubbles to start.
3) After bubbles have started to form, light the end of a wooden coffee stir stick, then blow it out.
4) Right after blowing it out, put the smoldering wood down into the jar or test tube, near the liquid but not touching it.
5) The wooden stick should start burning again.

What is happening:
The yeast is making an enzyme molecule that has the ability to tear that extra oxygen off the hydrogen peroxide, producing H_2O, water, and O_2, oxygen. Oxygen is necessary for combustion and encourages burning.   In the presence of oxygen, a recently extinguished flame will light again. Scientists in Lavoisier's day used wooden splints much like our wooden coffee sticks. Both Priestly and Lavoisier would have done this test many times.

4) LAB DEMO: Make "Elephant's Toothpaste" with hydrogen peroxide and yeast

This lab is often called "elephant's toothpaste" because it vaguely resembles a stream of toothpaste coming out of a tube, but on a giant scale. You can Google it or type it into YouTube and see many web pages and videos devoted to it. Feel free to check with other sources and use those instructions instead of what is printed here.

You will need: 1/2 cup hydrogen peroxide (6%), packet of yeast, water, 2-liter plastic soda bottle, dish detergent
        OPTIONAL: food coloring,

What to do:
1) Pour the hydrogen peroxide into the bottle.
2) Add a few drops of food coloring if you want your toothpaste to be colored.
3) Add a spoon of dish detergent. Swirl it around so the detergent is dissolved into the water.
4) In a separate dish, mix the packet of yeast with a few spoons of water.
5) Make sure your bottle is sitting somewhere that can get VERY messy, like a sink or the grass outside.
6) Pour the yeast water into the bottle and watch what happens.

What is happening:
The solution in the bottle will begin to foam, and then it will foam so fast that the foam will come shooting up out of the bottle, making a tall column until gravity collapses it. As in demo (3), the yeast is making an enzyme that pulls the extra oxygen off the peroxide molecule. The bubbles in the foam are mostly made of the oxygen gas that is being released.

5) LAB DEMO: Observing the products of combustion

In this lab, you will observe something else that those famous scientists saw, but did not understand. The appearance of water droplets during combustion was a profound mystery. They could not understand where the water was coming from. You have the advantage of living in an era where the chemistry of air and combustion is very well understood, so you can know exactly where those water droplets are coming from.

You will need: a small candle, a candle holder or lump of clay, a very clear glass jar   Optional: ice cube

What to do:
1) Light the candle.
2) Hold the bottom of the jar over the flame, but not so close that you get soot forming. We can look at soot in a minute. Right now you want to observe condensation of water droplets. The condensation will remind you of the "fog" that forms on a cold window when you breathe on it.
NOTE: If you have trouble seeing condensation, you can try cooling the glass by putting an ice cube inside the jar.
3) Once you have seen the condensation, go ahead and hold the jar closer and observe soot forming.

What is happening:
The condensation is made of tiny water droplets, H_2O. If you look at the picture of the wax molecule you can see where the hydrogens are coming from. The oxygens are coming from the air, of course. Carbon dioxide, CO_2, is also a by-product of combustion, and that's where the carbon atoms from the wax end up. The oxygens atoms in CO_2 are also coming from the air. When combustion works perfectly, all the wax atoms are converted into either water or carbon dioxide.
When you lowered the jar, you began restricting the amount of oxygen available for combustion. Not all the carbons could be made into H_2O or CO_2. Soot is made of carbon atoms that didn't join water or carbon dioxide molecules. INTERESTING SIDE NOTE: If you could look at the soot with a very high power electron microscope you would probably see some buckyballs. Soot is a very good source of buckyballs.

A WAX MOLECULE

6) LAB DEMO: Observing one way that carbon can be helpful

You will need: two small containers with lids, onion or garlic or something else smelly, water, charcoal (Horticultural charcoal from a garden shop works well. Or you might want to try foot odor pads.)

What to do:
1) Put a quarter cup or so of water in each container.
2) Put a quarter teaspoon of the smelly stuff into the water in each container.
3) Put as much charcoal as you can into just one of the containers.
4) Put the lids on and let them sit a while (15-20 minutes is probably enough).
5) Now take the lids off and sniff each container. Is there a difference? Does the container with the charcoal smell as strong as the one without?

What is happening:
Carbon compounds have the ability to absorb small molecules. That means that they grab onto the molecules and don't let go. The charcoal in your solution grabbed most of the molecules that would have otherwise gone off and floated into the air. It is these tiny molecules in the air that go into your nose and cause a smell. Carbon is used in many kinds of filters because of this ability to grab molecules.

7) AN ON-LINE REVIEW GAME

Here's an interactive website where you can play basic quiz games about the elements and the Periodic Table: **http://www.sheppardsoftware.com/Elementsgames.htm**

8) SKIT: "The Amazing Ramsay's Cryo-Show" (the discovery of noble gases)

If you don't want to bother with actually performing this skit, you can do it as "Reader's Theatre," with readers taking turns being Ramsay (or simply have the students read the skit to themselves).

Ramsay is the only actor in this skit, but if you want to perform it, Ramsay will need an assistant who will be behind the scenes (under the table) to make sure the props function at the right times.

The main character in this skit is Sir William Ramsay, the co-discoverer of krypton, neon, and xenon. He discovered all three in just three months—a chemistry world record. The year was 1898 and the months were June, July, and August. The discoveries were made possible by the invention of high-power refrigeration that could produce temperatures as low as -250 degrees below zero Celsius.

This skit is a humorous presentation of Sir Ramsay, and is not intended to be an accurate portrayal of his personality. The point of the skit is to learn about distillation of air, and how this technique can be used to separate the components of air. Distillation works because each element or substance has its own unique boiling/condensation point. Every substance can exist as solid, liquid, or gas, depending on the temperature and pressure around it. If the air around us is chilled sufficiently, its gases condense down into liquids. As the temperature is brought back up again, each substance will return to its gaseous state at a different time, allowing just that gas to be captured.

If you would like to perform this skit, I recommend the following props:

1) A fictional refrigeration unit (Use your imagination. It just needs to have a compartment that you can open and shut, and a thermometer that can be operated from the inside. Design the unit after you have read through the script.)

2) At least two large, identical containers that will fit inside the compartment in the refrigeration unit. One container will have a small amount of water in the bottom.

3) Protective gloves or hot pads (for imaginary use)

4) A visual aid about the three phases of water: solid, liquid, gas

5) A sign with the names krypton, neon, and xenon written on it

6) Optional: a magician's top hat for Ramsay

9) SKIT: "The Elusive Phlogiston" (the discovery of oxygen)

The cast list and instructions are on the first page of the script.

The phlogiston theory was an attempt to explain combustion. Before the discovery of oxygen, fire and combustion were a profound mystery. Phlogiston was a fictional "element" that was found in everything that burned. When the phlogiston was gone, the material stopped burning. The theory was accepted by all scientists for about 150 years. Cavendish missed discovering hydrogen because of his belief in phlogiston, and Priestly missed getting credit for the discovery of oxygen he thought it was phlogiston. It is important to remember that even theories that are accepted by most scientists can still be wrong. Lavoisier was able to disprove phlogiston because of his ability to weigh and measure elements so accurately. His scales revealed that phlogiston did not exist.

"The Amazing Ramsay's Cryo Show"
A skit about the distillation of air

Hello, ladies and gentlemen! Welcome to the Amazing Ramsay's Cryo- Show. No! No!, Don't cry! I didn't say "cry," I said "cryo." Cryo means very cold. Really, really, really cold. My refrigeration unit here makes the coldest day in January look like a tropical vacation!

Tonight, with the help of my refrigeration unit, I am going to make liquid appear out of thin air! I have here a container of ordinary air. I need a volunteer from the audience.

Volunteer, will you please inspect my container of air to make sure that it is really just air? Does it smell like air? Does it feel like air? Does it breathe like air? Thank you, (sir/ madam/miss). Our volunteer has confirmed that this container does, indeed, contain just normal air.

Now it is time for my magic! *(He puts the container into the refrigeration unit.)*

Here we go! *(Press a button or something.)* It will take just a minute or two for the temperature to get cold enough. While the temperature is dropping, let me remind you that all substances can exist in three forms: solid, liquid and gas. *(Show the visual aid, if you are performing. Also, if you are performing, the hidden accomplice should now be slowly dropping the temperature on the thermometer. Also, the assistant will need to switch the jars, replacing the empty one with the one that has a little water in the bottom.)* We usually think of water as a liquid, right? But if you heat it on the stove, it can turn into steam and go into the air. If you cool water in the freezer, it will turn into a solid. Well, the same holds true for any substance in the universe. Take oxygen, for example. We think of it as a gas that we breathe in. But if you cool it enough, oxygen will turn into a liquid!

Ah-- I see my machine is ready! It has reached minus 273 degrees below zero on the Centigrade scale. That's as cold as cold ever gets. That's absolute zero. It's impossible to go any colder!

Let's look at our container of air. *(Take container out of machine using protective gloves, for effect.)* Voila! All the gases that were in that air have turned into liquid! This, ladies and gentlemen, is liquid air!

Now, for the second part of my trick. I will now let the temperature rise very slowly. *(Puts the container back in.)* One at a time, the liquid gases will return to their gaseous state. *(The assistant should be sliding the thermometer up to -269.)* We will stop the temperature at minus 269 and see what has happened in the jar. *(Gets the container back out.)*

In the top part of this container there is now pure helium gas. I will now allow the helium to escape. There. Now back into the freezer. *(Assistant raises the thermometer to -253.)*

The temperature is now minus 253. Let's take a look again. *(Gets out container again.)* Now there is pure hydrogen in the container! We will let it escape and return the container

Minus 229. Now what do we have? *(Gets container back out.)* Presto! Pure neon!

(This pattern continues, with the resulting dialog.)
Up to minus 150. Pure nitrogen!

Up to minus 122.   Pure argon!

Up to minus 118. Pure oxygen!

Up to minus 64. Pure krypton!

Up past zero.   Pure xenon!

And finally back up to room temperature again.

(Assistant has switched jars again, so that Ramsay takes out the jar with no liquid.)

Whew! All that stuff was in the regular air all around us!

I can proudly say that I was the one who discovered three of these gases: krypton, neon, and xenon. I used the name "krypton" because "kryptos" means "hidden" and I thought this gas seemed hidden in with all the others we already knew about.   "Neon" means "new" and "xenon" means "strange." I guess I like mysterious-sounding names.

I think these newly discovered gases will be good for something someday, but right now I don't have a clue what. They don't react with anything. They just sit and do nothing. They are totally inert. What good is an inert gas?

Well, next time you take a breath of air, just think of the Amazing Ramsay's Cryo- Show. My discoveries will always be with you, floating around in the air nearby!

Thank you for coming, ladies and gentlemen.

"The Elusive Phlogiston"
A skit about the discovery of oxygen

NOTE: You can pronounce phlogiston as "FLODGE-i-stohn" or "flah-GIST-on." People who speak American English tend to use the first one, and people who speak British English tend to use the second one. You can use whichever you prefer.

Props you will need: candle, matches, clear jar, piece of wood, small notebook and pencil (for Lavoisier), optional test tube or small jar or red powder for Priestly, scales for Lavoisier if you have some, plus anything you want to add to give a hint of the historical period in which each scent is set. (Small table and chairs for Priestly and Lavoisier to sit at, with perhaps a fancy teapot and cups, etc.)
NOTE: You might want to use large "cue cards" with the dates and places written on them so you can show the audience. Or, you can have the narrator come on and say the place and time.

Cast:
-Narrator (can also hold up the cue cards between scenes if you are using them)
-Johann Becher, *(Yo-han Beck-er),* an alchemist
-Georg Stahl *(Gay-org Stall),* a student of Becher
- J. H. Pott, a student of Stahl (It was not possible to find out what J. H. stands for.)
-Johann Juncker *(Yo-han Yung-ker),* a student of Pott
-Joseph Priestly
-Antoine Lavoisier *(An-twon La-vwah-zee-ay)*
-Audience member 1
-Audience member 2

SCENE 1: Germany, 1669 (Hold up a cue card with this printed on it, if you are using cue cards)

Becher: Stahl, I think I have finally solved the mystery of fire!

Stahl: That's wonderful, Master Becher. I am so fortunate to be your student!

Becher: Yes, you are. You and I will go down in history as the people who discovered fire.

Stahl: I think fire has already been discovered.

Becher: You know what I mean. We are the ones who will unravel the deep mysteries about fire! You see, Stahl, the ancients thought that everything was made of four elements: water, earth, air and fire. But they had it all wrong. Everything is made of earth, but there are three kinds of earth. One of them, which I call "terra pinguis" is oily and catches fire easily. Substances that burn have a lot of terra pinguis in them.

Stahl: Yes, that makes sense. Things that burn contain a lot of the element that burns. So plants contain a lot of this element. But what about metals? I've seen you melt metals in your crucible.

Becher: Metals contain a very small amount of terra pinguis. Just enough to let them melt.

Stahl: So what does terra pinguis look like? Can we collect a bottle of it?

Becher: No, I don't think so. Terra pinguis is an element you never actually see.

149

Stahl: So how do you know it is there?

Becher: Well, it must be there. It must! How else could things burn?

Stahl: Yes, I see your point. I'll have to begin studying these three new elements.

Becher: I expect great things from you, Stahl. Some day, you'll be teaching your own students about terra pinguis and how it creates fire.

SCENE 2: Germany, 1703 (Stahl is now a teacher. Mr. Pott is his student.)

Stahl: So you see, students, every substance that burns contains an element than burns. My teacher called it "terra pinguis" but I prefer to call it... "phlogiston."

Pott: Floggee what?

Stahl: Phlogiston. It's from a Greek word meaning "flames," because phlogiston is very combustible. Fire is the visible evidence that phlogiston is leaving a substance. When all the phlogiston is gone, the fire stops and you are left with nothing but ashes. So, young Mr. Pott, if phlogiston left the wood and you ended up ashes, then what is wood made of?

Pott: Um...ashes and phlogiston?

Stahl: Correct! What a brilliant pupil you are! Yes, as long as the phlogiston remains in the wood you can't see the ash. When the phlogiston leaves, then you can see the ash that was there all along. Wood is made of phlogiston and ash.

Pott: Where does the phlogiston go?

Stahl: Into the air all around us.

Pott: But we breathe air.

Stahl: Yes, we do, but we breathe it back out again.

Pott: Yes, of course. So we shouldn't hold our breath or we'll catch on fire, right?

Stahl: Fortunately, you can't hold your breath long enough to let the phlogiston settle in.

Pott: Then...is that how dragons breathed fire? By holding their breath?

Stahl: Mr. Pott, you're brilliant! You shall follow in my footsteps as the next great alchemist. It won't be long until you are a teacher with your own students.

SCENE 3: Germany, 1740 (J. H. Pott is now a professor and is an expert on phlogiston.)

Pott: Today, students, I shall demonstrate to you the element phlogiston.
(Pott has a candle in a candle holder, set on a plate, and a clear glass jar taller than the candle.)

150

This candle contains both phlogiston and wax. By lighting the candle, I can drive the phlogiston out of the candle. *(He lights the candle.)* By putting the candle under a jar I can contain the phlogiston. *(He puts the jar over the candle.)* The air can only hold a certain amount of phlogiston. When the air becomes saturated with phlogiston, the candle will stop burning because the phlogiston can no longer leave the candle. *(He watches and waits until the candle goes out.)* There—the air is now is what I call "phlogisticated," meaning it is full of phlogiston and cannot hold any more. However, if I lift the jar and let the phlogiston out, then the candle can burn again in the fresh, "de-phlogisticated" air.

Juncker: Does phlogiston itself ever burn?

Pott: No, it cannot be consumed by fire.

Juncker: What does phlogiston look like?

Pott: We don't know exactly, but I can tell you that it definitely consists of a circular movement about its axis.

Juncker: *(perhaps looking a bit confused)* Hmm. Do we know anything else about phlogiston?

Pott: It is responsible for producing colors. And also, it starts fermentation, the process by which sauerkraut and wine are made.

Juncker: I've heard some scientists say that when they burn metals they get heavier, not lighter. Shouldn't something become lighter after the phlogiston goes out of it? How do you account for things weighing less after they burn?

Pott: That's simple enough. You see, in some metals, once phlogiston leaves, the remaining particles get more compact, so the metal weighs less.

Juncker: But I've heard reports that sometimes the burned metal actually gets a little larger. And it weighs more, too.

Pott: Well, in these cases, the phlogiston can weigh <u>less</u> than nothing. When negative-weight phlogiston is in a substance it will actually weigh less than it normally would. Once the phlogiston is gone, the substance will return to its normal weight, making it look like an increase in weight.

Juncker: Negative weight...I see. Only sometimes phlogiston has positive weight, too.

Pott: Much research remains to be done! Perhaps you will take up the challenge.

Juncker: And so I shall!

Narrator: Johann Juncker did indeed become a teacher of phlogiston. And he totally believed that phlogiston could have negative weight. He called it "levity."
   By this time, the theory of phlogiston was spreading all over Europe. In England, a scientist named Henry Cavendish managed to isolate what we now call the element hydrogen. But, alas— he thought it was the elusive phlogiston, so he never received credit for discovering hydrogen.

In Germany, an apothecary named Carl Scheele found a way to make pure oxygen gas. But, believing in the phlogiston theory, he thought he had isolated "de-phlogisticated air" and never received credit for discovering oxygen.

Then, in England, a minister and amateur chemist named Joseph Priestly figured out a way to collect pure oxygen. And, believing in phlogiston, he also thought he had isolated de-phlogisticated air. Now, Priestly happened to work as a tutor for a wealthy English family who often traveled to Europe. On once occasion, they took Priestly along with them. One evening, they went to a social gathering of French intellectuals, and Priestly just happened to be seated next to one of France's most brilliant chemists, Antoine Lavoisier.

SCENE 4: Paris, France, 1774

Priestly and Lavoisier are sitting together at a table.

Lavoisier: Now, tell me again how you were able to gather this de-phlogisticated air.

Priestly: I started with a red powder called mercuric calx. (*Lavoisier pulls out a notebook and pencil and begins taking notes.*) (*Optional: Have Priestly pull out a test tube of red powder (turmeric or paprika?) and say, "Here, I've brought some along with me.")* After pouring liquid mercury into the top of the tube, I then turned it upside down and submerged it into a bath of liquid mercury. This would allow me to catch any gases that resulted from heating the red powder.

Lavoisier: (*scribbling notes*) Yes, go on. Then what happened?

Priestly: So there I am, holding an upside down tube filled with my red calx. (*If you are using a test tube, he can show this with the tube.*)

Lavoisier: With liquid mercury at the bottom.

Priestly: Yes. Then I heat the red powder to a very high temperature. After several minutes, I begin noticing that the mercury level in the tube is going down. A gas is appearing in the tube. After about an hour I carefully remove the tube so that the gas does not escape.

Lavoisier: What is the gas like?

Priestly: If I blow out a candle, then put it into the tube, the candle suddenly relights again! Just like that, the flame comes back! It must have been de-phlogisticated and the candle introduced new phlogiston into the test tube.

Lavoisier: Anything else? (*still taking notes*)

Priestly: Yes, if I put a mouse into a jar with this gas, the mouse can survive for a very long time.

Lavoisier: Did you ever breathe it yourself?

Priestly: Yes, I did sneak a few breaths myself. It made me feel all light and airy, very full of life.

152

Lavoisier: Thank you, Mr. Priestly, for sharing this outstanding discovery with me. I am working with gases right now, myself, and I shall try this experiment in my own lab. Only, I think I'll try your experiment in reverse.

SCENE 5:   Paris, France, 1777, at a gathering of scientists

(All other actors can be other members of Lavoisier's audience.)

Lavoisier: I am here to announce the discovery of a new element! It is a gaseous element, and I have named it "oxygen" using the Greek word "oxy" meaning "acid." This new element, I believe, is required to form acids, so I've called it the "acid-maker."

Audience member 1: Where did you find this element?

Lavoisier: It's everywhere, all around us! It's part of the air. Mind you, it's only part of the air. I believe that air is not one element but a mixture of elements.

Audience member 2: How much of the air is this new element, oxygen?

Lavoisier: I estimate that one fifth of the air is oxygen. I know this because my experiments use very precise measurements. I've seen one exactly one fifth of the air disappear and then reappear in the metal I am burning.

Audience member 1: How can air go into metal?

Lavoisier: It sounds strange, but it's true. Particles leave the air and are joined to metal. This is why metal can weigh more after burning than before.

Audience member 2: But I thought that was due to the negative weight of phlogiston.

Lavoisier: Phlogiston? My scales know nothing of phlogiston! My scales tell me that the air got lighter by precisely the same amount that the metal got heavier. There is no such thing as phlogiston!

Audience member 1: But phlogiston is a well-established theory. It's been around for hundreds of years! It must be true!

Lavoisier: For thousands of years, people believed in the four elements: fire, water, air and earth. And that turned out not to be true. I think scientists in the future should base their theories not on what people in the past have believed, but on what their measurements tell them.

Narrator: And that was the end of phlogiston. What became of Lavoisier? He went on to make many more contributions to the science of chemistry. Then... the French Revolution began and he became a victim in that horrific reign of terror. The revolutionaries executed Lavoisier because he had too many rich friends. Today, France holds dear the memory of Lavoisier. They establed the Lavoisier Medal in his honor. They are proud of all his many inventions and discoveries, including his discovery of the element oxygen.

153

ACTIVITY IDEAS FOR CHAPTER 7

1) LAB EXPERIMENT: What happens when oxygen combines with iron?

We can't actually see with our eyes atoms rearranging and combining with different atoms. We can only see the results: changes in color, odor, temperature, texture, etc. In this experiment, the change we will observe will be a change in temperature, first, and then in color and texture.

Iron has the unfortunate property of combining easily with oxygen and water. The result is iron oxide, FeO(OH), rust. The oxygen atoms are looking to gain two extra electrons to fill their outer shells and the iron has two it wouldn't mind giving up, so they bond together. (Remember that we mentioned the fact that transition metals usually have several "oxidation states." Here is an example: sometimes iron and oxygen combine in a different way and form Fe_2O_3, which is a mineral called hematite.)

PART 1: Observing the exothermic nature of oxidation (allow 15-20 minutes)

You will need: a piece of steel wool (not the kitchen kind—the workshop kind), vinegar, a jar, a thermometer, a rubber band and a paper towel

What to do:
  1) Put the thermometer in the jar, screw on the lid, and let it sit for a few minutes. Record the temperature.
  2) Soak the steel wool in vinegar for a few minutes, then drain off excess vinegar and rinse with water. Pat with paper towel, but don't dry it completely. (The vinegar acts as a cleaning agent and gets rid of any coating on the steel wool that might prevent the oxygen from coming into contact with the iron.)
  3) Put the steel wool around the bulb at the bottom of the thermometer and secure with rubber band if necessary.
  4) Observe the thermometer for 10 minutes and record the temperature at 2 minute intervals.

What you will see: This reaction (oxygen combining with iron) is "exothermic" and releases heat. You will see the thermometer rise several degrees.

PART 2: Observing the "recipe" for rust (allow several hours, or overnight)

Rust is the informal name for iron oxide. As the name suggests, iron oxide is made of iron and oxygen. Oxygen from the air combines with the iron in the metal. However, the oxygen floating around in the air is O_2, two atoms of oxygen joined with a covalent bond. This bond is strong enough to keep the oxygens together under normal circumstances. If the bond was weaker, we'd have dangerous single oxygens floating around in the air, which would not be good. The bond is strong, but not so strong that it can't be broken when it needs to be, like when our bodies need to use the oxygen atoms in our cells. So the bond is the perfect strength: not too strong, not too weak.

O_2 in the air won't react with the iron unless water is present. As we know, a metal object can sit in a dry room for years and not rust. Water molecules are needed to facilitate the breaking of the covalent bond between the oxygens and the transfer of electrons between the iron atoms and the oxygen atoms.

The "recipe" for making rust is: IRON + O_2 (in air) + H_2O → RUST. You need all three things to make rust. So if we remove one of the "ingredients" the recipe should not work. In this experiment, you will remove some ingredients and see what happens.

You will need: steel wool, four small clear containers with lids, water, paper towels, and soap.
Optional: salt and/or baking soda

NOTE: I find this experiment a little tricky to get right (though I know it can be done). On one occasion, the jar of water turned rusty brown, which is not supposed to happen. I did not have time to track down what the issue was, but thought I should add this note in case you want to pour a little vegetable oil on top to prevent any gas exchange at the surface. (I've seen the vegetable oil trick mentioned on some web sites.)

What to do:

1) Put a piece of dry steel wool (golf ball size is fine) in one jar and put the lid on. This takes water out of the equation.

2) In another jar, put a piece of dry steel wool and a puddle of water in the bottom. Put the lid on. This represents the full "recipe" with both water and oxygen. (It's up to you whether to rinse the steel wool beforehand in order to get any coating off. I've had it work either way.)

3) Put a piece of steel wool in another jar, fill to the brim with water, then screw the lid on. This will be our way to try to keep oxygen out of the equation, though water does contain some dissolved oxygen. (Fish can breathe water so it must contain some oxygen.) However, it will be getting less oxygen than if was being exposed to air. (If we had a special vacuum apparatus designed to suck all the air out of an unbreakable jar, that would be better. But since we don't have one, we'll just have to make do the best we can.)

4) Rub soap into a piece of steel wool and then put it into a jar with a puddle of water. In this one you have all the ingredients present, but you are not letting the air (oxygen) come into contact with the iron. The soap is "hydrophobic" (hates water) and will not let any water molecules stick. (One option would be to skip this jar and do the baking soda water one instead.)

6) Optional: You might want to add two other jars: one with salt water all the way to the brim, and one with baking soda water, also to the brim.

7) Make some predictions about what will happen. Then let the jars sit for several hours or overnight.

What you should see:

Hopefully, there will not be any rust in the jars that have something missing from the recipe. If this is not the case, you might want to guess what may have happened. (Scientists in the real world often deal with surprising results. They might have to tweak their experiment several times before they get results they are satisfied with. Running experiments repeatedly and always getting the same result is part of good scientific method.) Salt should speed up the rusting process. Baking soda will prevent rusting because alkali solutions slow down the rusting process.

2) LAB DEMO: Magnetic breakfast cereal

Some breakfast cereals have iron added to them. Iron is magnetic, so would the cereals then be magnetic? Try it and see. The best cereal to try is Total® bran flakes, but you could try anything that has 100% RDA iron. (If the cereal has not had iron added to it, these experiments will not work.)

You will need: a box of high-iron breakfast cereal, a blender (if possible), a strong magnet, a zip-lock plastic bag

PART 1: Floating a flake

If you float a flake on top of a bowl of water and hold a strong magnet next to it, should move slightly. Make sure your magnet is strong enough. Neodymium is best, if you have one, but you can also get very strong iron alloy magnets.

PART 2: Mush in a bag

1) Put the cereal into a blender with some hot water. Puree until you have a smooth, watery mixture, then pour into a "zip lock" bag. (Small freezer bags work well.) Make sure the mixture is watery enough that it will slosh around in the bag.

2) Put a magnet on the outside of the bag and hold the magnet in place while you squeeze the bag. You want to allow every part of the mixture to flow past the magnet so that the magnet can attract as many iron particles as possible. TOP: Keep the magnet in one place on the bag—don't move it around.

3) Move the magnet away and look to see if there are black spots or lines in the cereal. These are the iron particles. You might even be able to get a black patch under the whole magnet if conditions are right (good cereal, good magnet, good technique).

3) LAB EXPERIMENT: Dissolving aluminum foil

This activity is only to be done with adult supervision. Sodium hydroxide, NaOH, is very caustic. In fact, everyone who is within range of the experiment should probably be wearing safety goggles and gloves.

NOTE: This activity generates hydrogen as a by-product. If you would like to try igniting small flammable hydrogen bubbles, here is a teacher's guide that suggests a way to do this with a few pieces of lab equipment (test tube, plug, hose, beaker): **http://www.chem.ed.ac.uk/sites/default/files/outreach/experiments/hydrogen-teach.pdf**

You will need: aluminum foil, glass dish, water, sodium hydroxide (NaOH) which is an ingredient in cleaning products designed to unclog drains (Buy the dry kind that comes as little crystals, and read the label to make sure it contains sodium hydroxide.)

What to do:
1) Put on your goggles and gloves.
2) Put some drain cleaner crystals into the glass dish. Add a little water and stir.
3) Put a piece of aluminum foil into this solution and watch what happens. (It might have a slow start. Aluminum always reacts with oxygen on its surface, creating a protective layer on the outside. Once the outer layer is gone the inner atoms will react more quickly.)

What is happening:
Once the reaction gets going, the foil should dissolve very quickly (and provide a few "Oooo!" moments).
The NaOH is being split up into Na^+ ions and OH^- ions. The water molecules are being split into H^+ and OH^- ions. The aluminum atoms are combining with OH^- ions to make $Al(OH)_3$, called aluminates. (If you remove an OH^- you'll get aluminum hydroxide, a substance that is used in antacid medicines.)
Here is a summary of what is happening (unbalanced equation):
Aluminum + NaOH + H_2O $\rightarrow$ Na^+ ions + H_2O + $Al(OH)_4$ + H_2 (The H_2 is hydrogen gas.)

4) ART PROJECT: Make a "loopy" 3D Periodic Table!

Print the following pages onto card stock. Cut and assemble according to instructions printed on the pages.

The idea for a table with loops for the d and f shells goes back almost a century. Then a man named Roy Alexander picked up the idea in the 1960s and made a preliminary design. In the 1990s he made a version to sell commercially and you can find his products at: allperiodictables.com.

Can you use these shapes to make a Periodic Table?

1) Write the atomic symbols and atomic numbers of the elements on the squares. The word "GLUE" should be right-side up as you work. In other words, you can use the word GLUE as your guide to make sure you don't have the rectangles upside down.

2) Cut out all three rectangles.

3) Fold the thinnest one into a loop and fold the end flaps back.

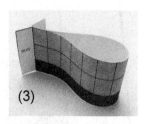

(3)

4) Cut the red line on the purple rectangle. You might want to trim out the whole red line (a strip about a millimeter wide) so that the fit won't be too tight when you insert the looped piece. (In other words, you want the red color to be completely gone.)

(5)

5) Insert the loop and glue in place.

6) Now make the purple rectangle into a loop, glue the end and fold back the flaps (the same thing you did to the first piece).

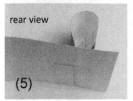

rear view

(5)

7) Cut the red line on the large rectangle. Trim out the whole red line (about a millimeter wide strip) to give enough space to insert the purple piece.

8) Insert the purple loop and glue in place.

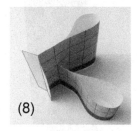

(8)

9) Bend the large rectangle into a cylinder and secure with glue on glue tab.

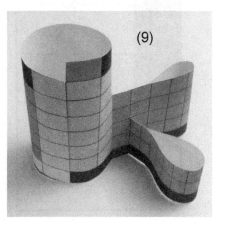

(9)

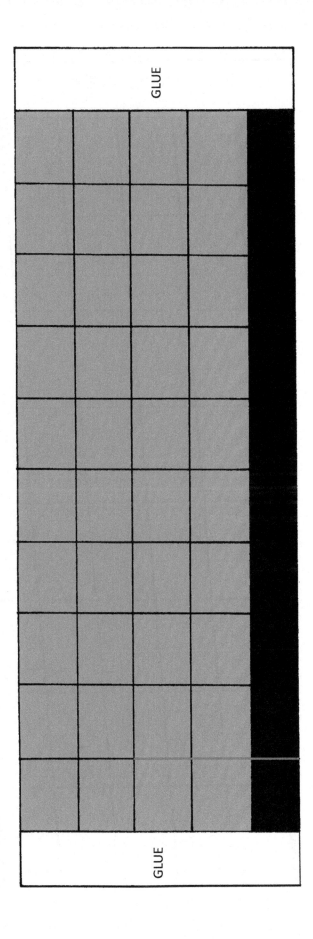

GLUE

GLUE

These patterns were developed completely from scratch by Ellen McHenry and have no connection to the official Alexander Arrangement. To see (or to purchase) the official model of Alexander's arrangement, visit allperiodictables.com.

COPY THIS PAGE ONTO CARD STOCK

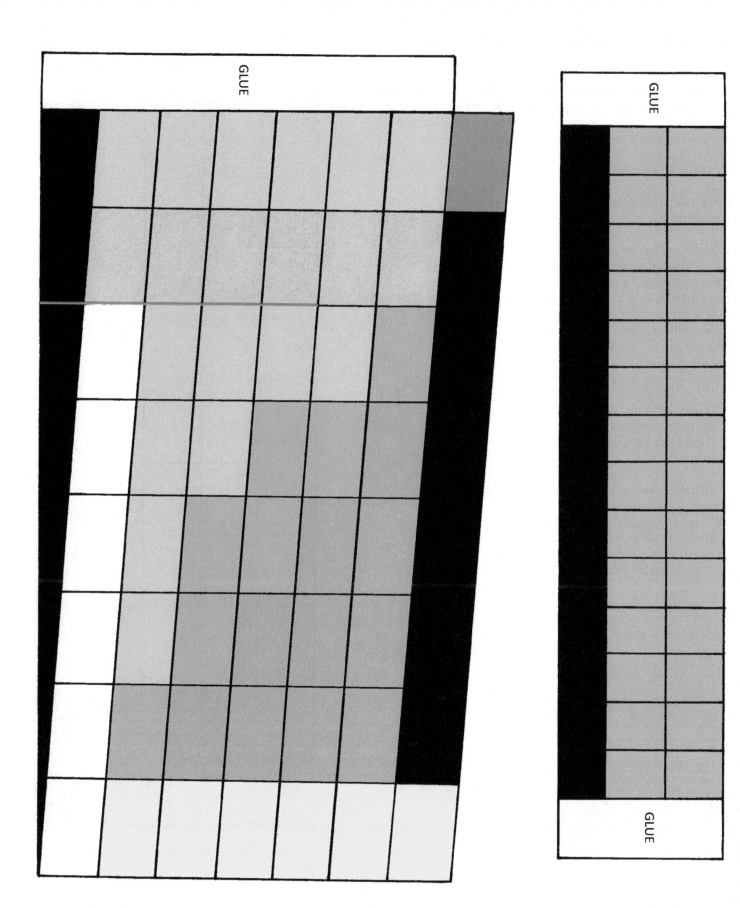

GLUE

GLUE

GLUE

These patterns were developed completely from scratch by Ellen McHenry and have no connection to the official Alexander Arrangement. To see (or to purchase) the official model of Alexander's arrangement, visit allperiodictables.com.

COPY THIS PAGE ONTO CARD STOCK   158

5) LAB EXPERIMENT: Build your own voltaic pile

This activity is only for those of you who <u>love</u> to do projects. It will take some time to round up the supplies, and you will have to do some online research (less than an hour, though) to be able to make decisions about what supplies to use. There are many variations of this projects, so search the Internet with key words "how to make your own voltaic pile." Some people recommend pennies and zinc-coated washers, others use nickels or aluminum foil. For the electrolyte, some instructions say to use salt water, and others will recommend vinegar or some other acidic solution. You will need to decide what you want to try. (Example site: **http://www.arborsci.com/cool/recreate-physics-history-build-a-voltaic-pile**)

NOTE: If you live in the USA, be aware that pennies made after 1982 contain very little copper. In 1982, the mints began making pennies out of 98% zinc with just a thin exterior coating of copper.

6) LAB EXPERIMENT: Cleaning pennies

How can you improve on something from a website called "copper.org?" I recommend visiting their site, learning more about copper, then trying their activity suggestion.

https://www.copper.org/education/Kids/copperandkids_cuexperiment.html

7) SKIT: "The Fame Game" A skit about the discovery of ruthenium

Once again, these skits do not represent an exact account of how these events happened. The overall story is accurate, but we don't have any historical transcripts telling us exactly what was said. I've taken the liberty of making up the dialogue.

Ruthenium was first discovered by Jedrzey Sniadecki (pronounced something like "Yed-jay Schnee-uh-det-ski") in 1807, but the Paris Commission wasn't impressed because he wasn't a famous scientist. Eventually he withdrew his claim and was forgotten. Then along came some scientists who were associated with big-name, well-known researchers and they had no trouble getting credit for their discoveries, even though two of those discoveries were later proven to be false. Sniadecki had been absolutely right, but he never got credit for his discovery. Life isn't fair sometimes, even in the world of science! Scientists are not immune from social politics.

If you would like to perform the skit, I recommend these props: a table and two chairs, books and papers to pile on the table, three sets of papers (first is a single sheet, second is a medium-sized stack, third is a thick stack), costume pieces for the commissioners, so you can age them between scenes: hat, glasses, white beard, mustache, etc.

You might want to use signs on the desk:

Paris, 1807
Several months later
Paris, 1828 (21 years later)
Paris, 1844 (another 16 years later)

8) SKIT: "Curie Finds a Cure" A skit about the discovery of polonium and radium

Yet another skit. Remember, you can use these simply as supplemental reading material—you don't have to perform them.   There is also a cartoon video presentation about Marie Curie on YouTube. This cartoon documentary is suggested as an activity in the student booklet. Simply go to YouTube and type in "Marie Curie part 1." The other parts will automatically pop up in the sidebar.

"The Fame Game"
A skit about the discovery of ruthenium

Characters:
- Jedrzej Sniadecki (Yed-jay Schee-uh-det-ski")
- Two members of the Paris Commission (which was an elite group of scientists)
- Wilhelm Osann
- Karl Klaus

SCENE 1: The office of the Paris Commission, 1807

The two members of the Paris Commission are sitting at a table, sorting through papers. A prominent sign on their desk says: "Paris, 1807"

First commissioner: What a busy day! So many people to interview! Seems like everyone is claiming to have discovered something nowadays. Who's next on the list?

Second commissioner: Mr. Jed-ra-zej Snee-a-deckee?

First: What? I mean, who? What kind of name is that?

Second: It says here he's from Poland, so I guess that would make his name Polish. I assumed YOU knew who he was.

First: Me? No. I've never heard of him.

Second: Just try to be polite anyway, okay? *(Calls out loudly:)* Mr. Sniadecki? Please come in.

First: You've come a long way, Mr. Snee-ad... Snee-ad jet-ski.

Sniadecki: Yed-jay Snchee-uh-det-ski, Sir. And, yes, sir, I've come all the way from Poland.

First: Hmmm... I don't know many scientists from your country. Well, tell us what you've got.

Sniadecki: Sirs, I have discovered a new chemical element. I have named it vestium. Here is my research that tells you how I found it. *(He puts a stack of papers on their desk.)*

The two commissioners slowly take the papers and flip through them, but look unimpressed.

First: Who do you work with?

Sniadecki: I work alone. But read my research. I'm sure the numbers are all correct. I know you will agree with me that I have discovered a new element.

Second: Have you ever worked with Mr. Wollaston or Mr. Tennant?

Sniadecki: No. As I said, I work alone. I've named the new element vestium, after the asteroid, Vestia, that was just discovered.

First: Okay, we'll check it out. Don't call us, we'll call you. Good day.

Sniadecki: Thank you for your time gentlemen. Please read...

Second: Next, please!

SCENE 2: Several months later

First: Mr. Sniadecki, come in.

Sniadecki: Thank you for seeing me again, gentlemen.

Second: Mr. Snee-ud...Shnee-u... Mr. "S," our scientists have tried to follow your procedure for extracting your new element but I am sorry to say that they cannot produce the same results you did. Are you sure you wrote down everything correctly?

Sniadecki: Yes, I am sure. I know it is an unusual method, but it has worked for me many times. I'm sure it is right.

First: Our scientists are the best in the world, you know. If they say there's no new element, then there's no new element. We're sorry.

Sniadecki: I'll take back my papers, then. I know my research is correct. But thank you for your time, gentlemen.

SCENE 3: Same office, 21 years later, in 1828

First: There's no end to interviewing people, is there?! Who's next?

Second: Mr. Wilhelm Osann. He claims he's discovered a new chemical element. Mr. Osann, come in, please.

Mr. Osann: Good morning, gentlemen. I have brought you my research demonstrating the existence of three new elements. I call them pluranium, ruthenium and polinium. *(He hands over a single sheet of paper.)*

The commissioners examine the paper.

First: Well, I see here that you've worked with Mr. Wollaston and Mr. Tennant. Very good!

Second: Anyone who works with Wollaston and Tennant must know what they are doing. We'll put you down as the discoverer of those elements. Congratulations!

Mr. Osann: Thank you gentlemen. Good day.

SCENE 4: Same office, 16 years later. (Paris, 1844)

First: Seems like we've been working here at this desk for almost forty years!

Second: We have. But someone's got to do this job. Who's next?

First: Mr. Karl Klaus.

Second: Mr. Klaus, come in, please!

Klaus: Good afternoon, gentlemen. I have come here today with evidence that two of the three elements supposedly discovered by Mr. Osann are not really elements at all. The elements pluranium and polinium don't exist! Only ruthenium is really an element. *(He hands a medium-sized stack of papers to them.)*

First: Hmm... I see you've worked with Mr. Wollaston and Mr. Tennant. Very good. Your research looks convincing. We'll remove those two other elements from our list and just leave ruthenium on.

Second: You've managed to improve the quality of the samples. Yours are much better than Mr. Osann's were.

First: Yes, in fact... your work reminds me of someone else's, a long time ago. Let me see if I can find that file.... Here it is. Look! Your work is almost exactly the same as the work done by this fellow over 35 years ago!

Second: Who was it?

First: It was a Mr. Schnee-ud... Schnee-uh-jet-ski?

Second: Sounds familiar, but I don't remember him. Too bad. Whoever he was, he was right after all!

First: Well, congratulations, Mr. Klaus. And give our regards to Mr. Wollaston and Mr. Tennant when you see them.

Klaus: I will, sir. Thank you, sirs. Good day.

"Curie Finds a Cure"
A skit about the discovery of polonium and radium

Cast:
-Narrator
-Marie Sklodowska Curie (*sklo-DOW-ska*)
-Broniswava Sklowdowska, Marie's older sister (the Polish spelling of her name is Bronisława)
-Mr. Sklodowska, Marie's father
-Kazimierz Zorawski, son of a wealthy Polish family (*KAZ-i-meerz*)
-Pierre Curie
-Nobel Prize committee member
-Irène Curie, their older daughter (*ee-REN*)

SCENE 1: Marie's childhood home in Warsaw, Poland, around 1880. Marie is about 10.

Broniswava: I don't think I'll ever be able to read these Russian textbooks. Father, why can't the Russians just leave our country? Why can't school be in Polish? There's nothing wrong with our language. It's just as good as Russian.

Mr. Sklodowska: My dear daughter, I am so sorry your life is so hard. Some day the Russians will leave and Poland will be free again. But right now our family has to do the best we can under the circumstances. And that means learning to read Russian so you can get a good education. You still want to be a doctor, don't you?

Broniswava: Yes, I do want to be a doctor, and that's the only reason I'm studying this stupid Russian grammar.

Marie: I'll help you with Russian, Broniswava, I started learning it at an earlier age than you did, so it was easier for me. I think I've got the grammar figured out.

Mr. Sklodowska: That's the way, girls-- help each other whenever you can.   With your mother gone now and our family's land and money having been taken from us by the Russians, the only way any of us will succeed is if we help each other every way we can.

Broniswava: It's probably pointless studying this grammar anyway, because we'll never be able to afford to send me to the university.

Marie: I know another way I can help you, Broniswava.

Broniswava: Yes?

Marie: I can find a job and work for a few years, and pay your university tuition. Then, when you are finished with school, you can work as a doctor and pay for my education. But I don't think I want to be a doctor. I want to study math and physics, like father did.

Broniswava: Marie, I'm not sure I want you giving me all the money you earn.

Marie: But it's like father said. We must find a way to help each other. Otherwise, neither of us will be able to study at the university.

SCENE 2: At the home of a rich Polish family, several years later

Narrator: When Broniswava was old enough to go off to college, it was arranged that Marie would work while Broniswava studied to be a doctor, then it would be Marie's turn to go to school. Broniswava went off to the Sorbonne, in Paris (and had to learn yet another language!) while Marie worked as a governess, teaching and taking care of the children of a rich family. Marie fell in love with the oldest son in the family, who was about her age.

Kazimierz: Marie, you know I love you, too, but my mother and father are against us getting married. They want me to marry a girl from a wealthy family, and your family is... well...

Marie: Poor. I know we don't have money, but we used to. It isn't fair. Just because my grandparents stood up against the Russians, they lost all their fortunes. It's not fair.

Kazimierz: I know it's not fair, Marie, but there's nothing I can do about it. I can't marry without my parents permission. They won't give me my inheritance if I marry you.

Marie: I must find another job, then. I can't stand staying here any longer, knowing we won't be able to marry. Good bye, Kazimierz.

SCENE 3: Paris

Narrator: So Marie left. Not long after this, she got a letter from Broniswava asking her to come to Paris. Broniswava had just gotten married and wanted Marie to come live with them. Broniswava was anxious for Marie to finally start her college education. When Marie arrived in Paris, she didn't know any French. But this didn't stop her from starting to attend math and physics classes right away.

Marie: Broniswava, do you remember when you used to complain about having to learn Russian?

Broniswava: Yes, of course I remember.

Marie: And here we are having to learn French, too!

Broniswava: But it was different this time, Marie. I chose to learn French. It wasn't forced on me like Russian was. I wanted to learn French so that I could attend the Sorbonne, one of the most prestigious universities in the world. I'm here because I want to be here.

Marie: At first I didn't want to come, Broniswava. I only came because you insisted. But now that I am here and I am learning everything I wanted to know about math and physics, I'm glad I came. I think I want to earn two degrees-- one in math AND one in physics.

Broniswava: That's a bit ambitious, Marie! One degree is enough for me!

SCENE 4: A laboratory in Paris

Narrator: Marie did indeed earn two degrees. After her degree in physics, she earned a degree in math only one year later. Immediately following graduation she began working in a laboratory studying something that fascinated her: magnetism. But she wasn't the only one fascinated with magnetism. Pierre Curie was equally fascinated, and he was also working at that lab.

Pierre: I am fascinated with the magnetic properties of this steel!

Marie: So am I. Could you pass the electrodes? Let's hook this up and see what results we get.

Pierre: Marie, you know what else fascinates me? You. I've never met anyone like you before. Maybe it's just the magnetism gone to my head, but will you marry me?

Marie: I didn't think I could ever love again after Kazimierz broke my heart, but yes, Pierre, I accept your offer of marriage. I think I love you, too.

Narrator: And so Marie and Pierre became not only lab partners, but life partners. They were hardly ever apart. Then Marie decided to start researching something that had just been discovered. A man named Henri Becquerel (*Beck-er-ell*) had demonstrated that uranium gives off invisible high-energy rays. Marie wanted to find out what these rays were, and where, exactly, they were coming from. In order to get the uranium she needed, she had to boil down a dirty, black mineral ore called pitchblende.

Pierre: Marie, that's a huge mound of pitchblende you've brought into the lab courtyard. You must have a plan. And what are you doing with my electrometer?

Marie: Pierre, this is very strange. The electrometer shows that the pitchblende is four times as electrically active as a sample of pure uranium. Logically, I must conclude that there is something else in this pitchblende, besides the uranium. A mystery element that is four times as active as uranium.

Pierre: That makes sense, Marie. I think you must be right.

Marie: I intend to boil down this big pile of pitchblende and try to extract that element from it.

Pierre: That's a pretty big pile. It could take months to boil down that much pitchblende!

Marie: Well, I'd better not waste time then! Do we have a very large kettle I can use for a couple of months?

Narrator: Marie stirred and boiled for weeks until she had produced a very small amount of what she was sure was a new element. She decided to name it "polonium," honoring her homeland of Poland. Marie and Pierre were awarded the Nobel Prize in Physics in 1903 for their discovery of polonium and their research on radioactivity. Marie was first woman to ever win this prize.

Nobel Prize committee member: Congratulations, Pierre and Marie. This award is for your outstanding research on radioactivity. We hope you will keep researching.

Narrator: They did keep researching.

Marie: Pierre, I'm convinced there is still another element hiding in the pitchblende.

Pierre: I think you are right, Marie. In fact, I am so sure of it that I am going to temporarily give up my study of crystals to help you find that second new element.

Narrator: So Marie and Pierre worked together day after day. Finally, after months of work, they managed to isolate enough of the new element to be able to announce to the world its discovery. They decided to call it "radium" after the Greek word "radius," meaning "ray." They also made up a new word to describe these elements that produced so much energy. They called them "radioactive." Unfortunately, Marie and Pierre did not know that radioactivity was dangerous.

Marie: Pierre, now that we have found this second new element, we must produce enough of it so we can test it to find out its properties.

Pierre: Yes, of course, Marie. I will continue to help you work on these new elements.

SCENE 5: Paris and Warsaw

Narrator: Then, one day, tragedy struck. Pierre was killed in a traffic accident. Marie was left as a widow with two young daughters to raise on her own. But she continued working on isolating radium. Her daughters grew up watching their mother's devotion to science and they became scientists themselves. They also saw her receive a second Nobel Prize in 1911.

Irène: I want to grow up to be a scientist, just like my mother. And I want to marry a scientist just like my dad. And maybe I'll win a prize, too.

Narrator: Irène did just that. The family tradition carried on. And speaking of family, Broniswava comes back into our story. Marie found out that radium could be of great use to doctors since it could produce X-rays. She helped World War I doctors by designing portable X-ray units that could be taken onto the battlefields. Radium also appeared to be very useful in the treatment of some cancers. Marie raised enough money to open the Radium Institute in Warsaw, Poland, and Broniswava was made the director.

Broniswava: Marie, look what we have done together, you and I! Look what sisters can do if they help each other.

Marie: Yes, Broniswava. We will be able to help so many people in Poland as well as around the world. My hope is that radium will help to cure many diseases.

Narrator: Though radium would cure many people in the future, it made Marie sick. After many years of handling highly radioactive substances with no protection, Marie died at age 60 of an illness caused by the radioactivity. She continued to receive honors, however, even after her death. A large statue of Marie was built in front of the Radium Institute. It is said that when Kazimierz was an old man, he would come and sit in front of the statue and stare at it.

Kazimierz: Marie, Marie, why didn't I marry you?!

ACTIVITY IDEAS FOR CHAPTER 8

1) GROUP GAME: "One of my elements is missing!"

You will need: the entire deck of Quick Six cards.

What to do:

      Lay out the cards on a table or on the floor so that they form the Periodic Table. Take turns being the "hider" and the "guesser." The guesser closes his eyes while the hider chooses one of the cards and removes it from the Periodic Table, hiding it behind his back. When the guesser opens his eyes, he tries to guess which element is missing. This game can also be played in a competitive fashion, with teams who try to be the first to correctly guess the missing element. (In my class, I had a Periodic Table shower curtain on the wall, and I used a black paper square the same size as the element squares, and used tape to make it stick.)

2) GROUP DISCUSSION: Naming a new element

      Ask the students to answer this question: If you were the discoverer of a new element, what would you call it? Would you name it after yourself? a famous scientist? a recent discovery? Someone from mythology?

3) SNACK: Eat a radioactive snack (yes, really!)

      We eat and drink small amounts of radiation every day. Our bodies are able to deal with this tiny amount of naturally-occurring radiation, so we don't need to be overly concerned about it. For example, potassium is one of the essential elements that our bodies need, so it's good for us to eat foods such as bananas, which are high in potassium. Most potassium atoms (93.3% of them) have 39 neutrons in their nucleus. The rest have either 40 or 41 neutrons. In the case of potassium 40, the extra neutron eventually turns into a proton, creating a ray of radioactivity (a beta particle). In an average-sized human body, about 4,000 potassium atoms decay every second. You also consume small amounts of radioactive elements such as uranium, thorium and radium. Remember, though, this is a naturally-occurring phenomenon, and should not discourage us from eating fruits and veggies!

      You might like to provide a snack featuring foods that are high in potassium, and thus are higher in natural radiation than other foods.   High-potassium foods include dried fruit (apricots, peaches, figs, dates, raisins), fresh bananas and cantaloupe, orange juice and grapefruit juice, potatoes, winter squash, all types of beans (including lentils), Brussels sprouts, tomatoes, carrots, zucchini, broccoli, and fish.

      If you would like to eat something containing radium, snack on Brazil nuts. Brazil nuts are 1,000 times more radioactive than any other food. The radium does not stay inside your body, though; it moves right on through and exits. If you want an exact figure for how many pCi/g's they contain, go to:

      http://www.orau.org/PTP/collection/consumer%20products/brazilnuts.htm

4) FINAL REVIEW GAME: "ELEMENT CONNECTIONS" (A bingo-type game)

      Use the pattern on page 171 to make the game "Element Connections." It is a bingo-type game. To get a "bingo" in this game you need to connect circles from the top bar to the bottom bar, but they don't need to be in a straight line—they may be in any order as long as they connect in a continuous string (no diagonals). For instance, your string may go: down, down, over, over, down, down, over, over, down, down, over, down. (Imagine that the lines and pennies can conduct electricity and think of making one, long, continuous electrical connection from top to bottom.) The circles are perfect for pennies, but you can use other tokens as well.

5) OBSERVATION: Fluorescence in household substances

You will need: a black light and various household substances that fluorescence, such as laundry soap, highlighter markers, petroleum jelly (Vaseline), white peppermint candies, white paper or envelopes, wicks of candles, some white craft supplies such as chenille stems and yarn (For a complete list, just Google "household items that fluoresce.)

       NOTE: You might also want to supply some things that do not fluoresce and predict which ones will and which ones won't.

       Fluorescence is when electrons have been excited up to a higher energy level, then fall back down to their original level, giving off the extra energy as (in this case) visible light. The energy going in is UV light (the kind that is just above violet in the rainbow, not the high-energy kind that damages skin). The energy coming out is visible light but at the blue/purple end of the rainbow. Your eyes interpret this as a whiter-than-white glow, which is why so many soaps and fabrics contain fluorescent dye.

6) MEMORIZATION CHALLENGE

You will need: a copy of the certificate for each student (you'll find the pattern page at the very end of this section), a blank Periodic Table (the pattern for the pillowcase is fine), and a supply of candy or treats of various sizes, ranging from small single pieces to full-size or jumbo-size bars (optional: plastic baggies for collecting treats) You can provide non-edible prizes, too, such as pencils, coins, stickers, etc.

How to set it up:
       Place the blank Periodic Table on the table. Just to the right of each noble gas, put a small pile of candy. The smallest treats (such as a single Hershey's Kiss, a piece of bubble gum, or a (wrapped) Lifesaver) go next to helium. Slightly larger treats (such as a stick of gum or a miniature chocolate bar) go to the right of neon. An even larger treat goes to the right of argon (a "snack size" bar?).   Continue this pattern, until you end up with a very large candy bar (the kind that is about six inches long!) next to radon.   Not everyone will make it to radon, so you won't have to buy "radon bars" for everyone. If you are working with a class, you might want to take a survey ahead of time to see how many students plan on memorizing to radon.

How the challenge works:
       Students take the challenge one at a time. If the student can successfully say the elements in the first row, he may pick up a treat at the end of the first row. (The first row is a guaranteed success, since all they have to say is: "hydrogen, helium." No one goes away empty-handed in this activity.) If the student can say the elements in the second row, he gets to pick up the slightly larger treat next to argon.   This continues, until the student wants to stop.
       Optional: For the elements in the last two rows, you may want to provide rewards for getting halfway across. Perhaps if they get halfway they can have the treat from the previous row? It's up to you.

How to use the certificate:
       Circle (or color the square of) the highest element that the student reached in their recitation. Sign and date the certificate, and give the student a hearty "Congratulations!" for their effort.

7) A FINAL REVIEW/TEST

       You can use the following pages either as a final exam or simply as a final review. Or you can skip them.

ELEMENT CONNECTIONS

Make a copy of the game board for each player and tell them to fill in the circles with symbols of elements on the main part of the table (no actinides or lanthanides). The symbols should be randomly placed so that each board is unique. Provide pennies, or other tokens, for the players to put on the circles as clues are called. When a player has tokens placed on a continuous pathway from top to bottom, he calls out "Connection!" The player must then read off the elements he has along the pathway and his answers are checked.

You may use these clues in any order. Just make sure you write down the answers on a slip of paper as you go long so you can check student answers easily and quickly.

Clues:

• This element has no neutrons. (hydrogen)
• This element has 48 protons. (cadmium)
• The name of this element means "stench." (bromine)
• This element is named after the town of Ytterby (it-er-bee), Sweden. (yttrium)
• This alkali earth metal burns red in fireworks. (strontium)
• This element has 18 electrons. (argon)
• This is the smallest element that is not a gas. (lithium)
• This element was first discovered in the sun. (helium)
• This alkali earth metal is used to take x-rays of the digestive system. (barium)
• You may mark a radioactive element of your choice. (technetium, polonium, astatine, radon)
• This alkali metal is very abundant in bananas. (potassium)
• This semi-metal is famous for its use as a poison. (arsenic)
• This element has 15 electrons. (phosphorus)
• This element has electron configuration $1s^2\ 2s^2\ 2p^6\ 3s^2$. (magnesium)
• This transition metal is used to repair bones. (titanium)
• This element has 9 protons. (fluorine)
• This element can be found occurring naturally as a light yellow mineral. (sulfur)
• This is highest element (highest atomic number) that is NOT radioactive. (bismuth)
• This element bonds with itself to form the hardest substance on earth. (carbon)
• This element is a gas with a valence of -3. (nitrogen)
• This element has a valence of -1 and used to be used for first aid. (iodine)
• The other name for this element is wolfram. (tungsten)
• Chemical compounds containing this element are often called "ferrous." (iron)
• This heavy, gray metal was once used to make water pipes. (lead)
• Don't confuse this element with magnesium! (manganese)
• This element has 30 electrons and is used to galvanize metals to prevent rust. (zinc)
• This is the only "happy" atom in the row that iron is in. (krypton)
• This element is in the same row as silver and the same column as nitrogen. (antimony)
• This element has a valence of 4 and was named after the Earth. (tellurium)
• This element has 49 protons. (indium)
• You may choose one element that has a valence of +1. (lithium, sodium, potassium, rubidium, cesium)
• This element combines with oxygen to make sand. (silicon)
• This element has an atomic weight of 16. (oxygen)
• This element was named after the Greek god Tantalus. (tantalum)
• This element has 5 neutrons. (beryllium) (You must subtract atomic # from atomic weight.)
• This element has electron configuration $1s^2\ 2s^2\ 2p^6$. (neon)
• This element is the only radioactive element in its row. (technetium)

• This alkali earth metal is found in bones and in concrete. (calcium)
• This is the most reactive, (but non-radioactive), member of the alkali metals. (cesium)
• This element has 44 protons. (ruthenium)
• The Latin name for this element is natrium. (sodium)
• The average weight of an atom of this element is about 190. (osmium)
• This is the lightest member of the true metals. (aluminum)
• This heavy transition metal is a liquid at room temperature. (mercury)
• This precious metal is current worth more per ounce than gold. (platinum)
• This element was named after Germany and is used in electronics. (germanium)
• You may choose one element that is a metal used in coins. (gold, silver, tin, zinc, copper, nickel)
• This element has a whole series named after it. (lanthanum)
• This shiny transition metal is used on vehicles because it is so resistant to corrosion. (chromium)
• This true metal is named after France. (gallium)
• This is the heaviest noble gas that is not radioactive. (xenon)
• Pewter is made mostly of this metal. (tin)
• This element was named after Marie Curie's homeland, Poland. (polonium)
• This element was named after Scandinavia. (scandium)
• This element has 23 electrons. (vanadium)
• This element was named after the asteroid Pallas. (palladium)
• This element has 72 protons. (hafnium)
• This element has a valence of -2 and is in the same row as potassium. (selenium)
• The average weight of an atom of this element is about 204. (thallium)
• The name of this element comes from the Latin word for rainbow: "iris." (iridium)
• This is the heaviest member of the halogen family. (astatine)
• This element has 42 protons. (molybdenum)
• This transition metal combines with O and Si to make a clear, diamond-like gemstone. (zirconium)
• The average weight of an atom of this element is about 93. (niobium)

ELEMENT CONNECTIONS

Fill in each circle with the symbol of an element. Use only the elements in the rows that begin with H, Li, Na, K, Rb, and Cs. Don't use any lanthanides or actinides.)

FINAL REVIEW

Name _____

(NOTE: You are allowed to look at a Periodic Table while doing this activity.)

ATOMIC SYMBOLS

Can you remember the symbols for these elements?

1) nitrogen ____   3) fluorine ____   5) chlorine ____   7) helium ____   9) carbon ____

2) gold ____   4) iron ____   6) magnesium ____   8) lithium ____   10) zinc ____

Can you remember the elements for these symbols?

11) Pb _____   13) Ag _____   15) Hg _____   17) K _____

12) Xe _____   15) Ar _____   16) Na _____   18) Al _____

QUESTIONS

19) The elements are listed in numerical order. Hydrogen is 1, helium is 2, lithium is 3, etc. What do the numbers mean?
   a) the mass (weight) of the atom   b) the number of protons it has   c) the number of neutrons it has
   d) the order in which it was discovered   e) the size of the atom

20) The word "valence" means:
   a) the number of protons an atom wants to eject from the nucleus
   b) the mass (weight) of an atom
   c) the number of electrons an atom wants to gain or get rid of
   d) whether an ion is positively or negatively charged

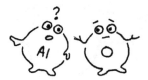

21) If you were doing an experiment in which you wanted electricity to flow through an element, which of these elements would you choose?   a) iodine   b) gold   c) sulfur   d) magnesium   e) mercury

22) What is the best and easiest way to separate the sodium and chlorine atoms in NaCl?
   a) smash NaCl with a hammer   c) put the NaCl into water
   b) put electricity through the NaCl   d) pull the atoms apart with tweezers

23) Which of these statements is NOT true about carbon?
   a) Carbon atoms can bond with <u>any</u> element on the Periodic Table.
   b) Carbon has a valence of +4 or -4.
   c) Carbon has the ability to grab small molecules and hold on to them.
   d) Carbon is one of the key elements in the chemistry of living things.
   e) Carbon is the element that diamonds are made of.

24) In which type of bonding do the atoms share their electrons?
   a) covalent   b) ionic   c) metallic   d) all of these

25) In which type of bonding are the electrons able to move about freely?
   a) covalent   b) ionic   c) metallic   d) all of these

TRUE or FALSE?   (Write T or F on the line.)

26) ____ Hydrogen atoms do not have any neutrons.

27) ____ The noble gases are toxic to breathe, just like chlorine gas is.

28) ____ The elements in the actinide series are the only radioactive elements on the Periodic Table.

29) ____ Ionic compounds (that join an alkali metal to a halogen) are called salts.

30) ____ If an atom were as large as a sports stadium, the nucleus would be about the size of a marble.

"ODD ONE OUT" Figure out which one in each set does not belong and circle it.
Consider the chemical properties of the elements, as well as what family groups they belong to. (Don't consider letters or numbers or where names came from, just chemical and physical properties.)

31) xenon   argon   krypton   oxygen
32) lithium   sodium   magnesium   potassium
33) technetium   rhodium   ruthenium   molybdenum
34) iridium   platinum   gold   mercury
35) carbon   iron   phosphorus   sulfur
36) europium   thorium   gadolinium   terbium
37) chlorine   neon   bromine   nitrogen
38) uranium   polonium   francium   californium
39) iron   tin   lead   bismuth
40) iron   aluminum   neodymium   samarium

Just think—
Humphry Davy
never owned a
Periodic Table!

MATCH THE ELEMENT WITH ITS DISCOVERER (a few of these are from the skits)
Use these as possible answers: oxygen, magnesium, ruthenium, iodine, radium

41) Marie Curie _____
42) Humphry Davy _____
43) Antoine Lavoisier _____
44) Bernard Curtois _____
45) Karl Klaus _____

Dmitri Mendeleyev didn't discover any elements.

VALENCIES   (Possible answers: -2   -1   0   +1   +2)

46) All the alkali metals have a valence of ____.
47) All the noble gases have a valence of ____.
48) All the alkali earth metals have a valence of ____.
49) All the halogens have a valence of ____.
50) Oxygen has a valency of ____.

FIND THE FAKES
        Consider each molecule carefully. Would the atoms really bond in this way? (For example, chlorine would never bond with fluorine because they are both -1. They each need to find some element that is +1.) Write YES if the molecule is possible, and NO if it is not.

51) KNa _____   53) KCl _____   55) AlGa _____   57) MgO _____   59) MgS_____
52) NaI _____   54) HeCl _____   56) HCl _____   58) CaF _____   60) LiF _____

A FEW MORE TRUE/FALSE QUESTIONS

61) ____ Elements that are in the same column (up and down) on the Periodic Table are likely to have similar chemical properties.
62) ____ Radioactivity is when outer shell electrons jump to a higher energy level then fall back down.
63) ____ Magnetism is caused by electrons all spinning in the same direction.
64) ____ Radon is the last (highest number) naturally occurring element on the table.
65) ____ Most elements on the table look like gray or silver metals when in their pure form and not combined with anything else.

ANSWER KEY for FINAL REVIEW

1) N   2) Au   3) F   4) Fe   5) Cl   6) Mg   7) He   8) Li   9) C   10) Zn

11) lead   12) xenon   13) silver   14) argon
15) mercury   16) sodium   17) potassium   18) aluminum

19) b   20) c   21) b   22) c   23) a   24) a   25) c

26) T   27) F   28) F   29) T   30) T

31) oxygen, because it is not a noble gas
32) magnesium, because it is not an alkali metal
33) technetium, because it is radioactive and the others are not (but they are all transition metals)
34) mercury, because it is a liquid at room temperature
35) iron, because it is not a non-metal like the others
36) thorium, because it is an actinide, not a lanthanide
37) bromine, because it is a liquid at room temperature and the others are gases
38) californium, because it is not naturally occurring and must be made in a lab
39) iron, because it is not a true metal
40) aluminum, because it is not magnetic

41) radium   42) magnesium   43) oxygen   44) iodine   45) ruthenium

46) +1   47) 0   48) +2   49) -1   50) -2

51) No   52) Yes   53) Yes   54) No   55) No   56) Yes   57) Yes   58) No   59) Yes   60) Yes

61) T   62) F   63) T   64) F   65) T

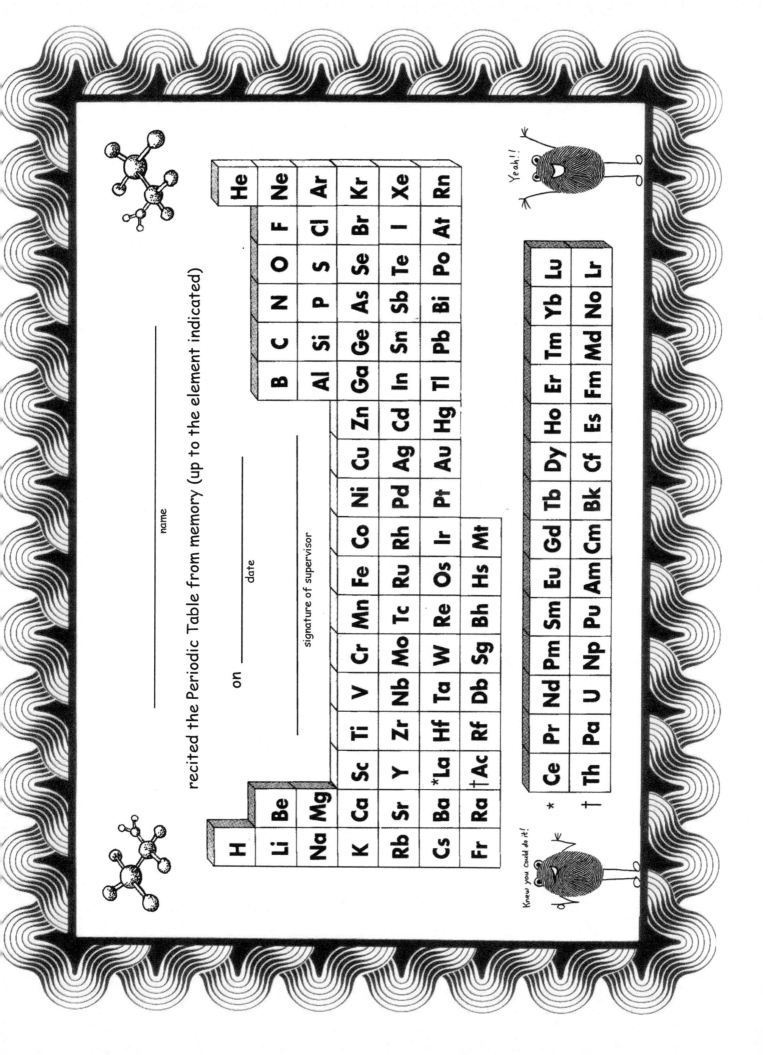

recited the Periodic Table from memory (up to the element indicated)

name

on

date

signature of supervisor

H																	He
Li	Be											B	C	N	O	F	Ne
Na	Mg											Al	Si	P	S	Cl	Ar
K	Ca	Sc	Ti	V	Cr	Mn	Fe	Co	Ni	Cu	Zn	Ga	Ge	As	Se	Br	Kr
Rb	Sr	Y	Zr	Nb	Mo	Tc	Ru	Rh	Pd	Ag	Cd	In	Sn	Sb	Te	I	Xe
Cs	Ba	*La	Hf	Ta	W	Re	Os	Ir	Pt	Au	Hg	Tl	Pb	Bi	Po	At	Rn
Fr	Ra	†Ac	Rf	Db	Sg	Bh	Hs	Mt									

*	Ce	Pr	Nd	Pm	Sm	Eu	Gd	Tb	Dy	Ho	Er	Tm	Yb	Lu
†	Th	Pa	U	Np	Pu	Am	Cm	Bk	Cf	Es	Fm	Md	No	Lr

Yeah!!

Knew you could do it!

BIBLIOGRAPHY

These are the books I used when I first wrote this, back in 2001. Since then the world has gone digital, and during the years since then, I've read many websites and watched a number of documentaries. Unfortunately, at first I didn't think about keeping track of all the virtual resources I consulted (like I do now). At the bottom I've listed some of the web addresses I've used to update the later chapters.

General Chemistry; Principles and Structure (Third Edition) by James E. Brady and Gerard E. Humiston.
Published by John Wiley & Sons, ©1982.

The Chemical Elements by I. Nechaev & Gerald Jenkins.
Published by Tarquin Publications, UK. © 1997. ISBN 1-899618-11-2

Exploring Chemical Elements and Their Compounds by David L. Heiserman.
Published by TAB Books, a division of McGraw-Hill, © 1992

Chemistry for Changing Times (8th edition) by John Hill and Doris Kolb
Published by Prentice Hall in 1998. ISBN 0-13-741786-1

The Visual Dictionary of Chemistry published by Dorling-Kindersley, ©1996

Chemistry For Every Kid by Janice Van Cleave. John Wiley & Sons, © 1989.

Descriptive Chemistry by Donald McQuarrie and Peter Rock.
Published by W. H. Freeman and Company, New York, © 1985. ISBN 0-7167-1706-9

Usborne Dictionary of Science published by Usborne, UK.

Eyewitness Handbook of Rocks and Minerals published by Dorling-Kindersley, © 1992

150 Captivating Chemistry Experiments Using Household Substances by Brian Rohrig.
Published by FizzBang Science, Plain City, Ohio, © 1997. ISBN 0-9718480-2-5

During the past two updates, I've read quite a few Wikipedia articles, plus articles on "howstuffworks.com."
Also not listed here are the many documentaries and video clips from YouTube. (A few of these appear in the playlist.)

Samples of the types of webpages I've consulted recently:

http://www.chemguide.co.uk/atoms/properties/3d4sproblem.html
http://www.chem4kids.com/files/elements/029_shells.html
http://www.rsc.org/eic/2013/11/aufbau-electron-configuration
http://scienceposse.blogspot.com/2011/01/rare-earth-question-what-do-f-orbitals.html
http://hyperphysics.phy-astr.gsu.edu/hbase/atomic/auger.html
http://www.namibiararearths.com/rare-earths-industry.asp
http://geology.com/articles/rare-earth-elements/
http://www.madsci.org/posts/archives/2001-01/980638580.Ch.r.html
http://www2.uni-siegen.de/~pci/versuche/english/v44-10.html
http://www.rsc.org/images/essay1_tcm18-17763.pdf
https://cosmosmagazine.com/physics/how-make-superheavy-element
http://highschoolenergy.acs.org/content/hsef/en/how-do-we-use-energy/combustion-and-burning.html

CPSIA information can be obtained
at www.ICGtesting.com
Printed in the USA
BVOW07s0246210517
484634BV00006B/17/P